Modern Distributed Tracing in .NET

A practical guide to observability and performance analysis
for microservices

Liudmila Molkova

BIRMINGHAM—MUMBAI

Modern Distributed Tracing in .NET

Group Product Manager: Kunal Sawant

Publishing Product Manager: Akash Sharma

Book Project Manager: Manisha Singh

Senior Editor: Rohit Singh

Technical Editor: Maran Fernandes

Copy Editor: Safis Editing

Proofreader: Safis Editing

Indexer: Subalakshmi Govindhan

Production Designer: Shankar Kalbhor

Developer Relations Marketing Executive: Sonia Chauhan

First published: June 2023

Production reference: 1160623

Published by Packt Publishing Ltd.
Livery Place
35 Livery Street
Birmingham
B3 2PB, UK.

ISBN 978-1-83763-613-6

www.packtpub.com

To Sasha and Nick, the constant source of inspiration and joy.

- Liudmila Molkova

Foreword

I have had the pleasure of knowing and working with Liudmila for many years. She is a visionary architect with a gift for practical implementation. This book is a testament to her unique combination of skills. If you want to get familiar with the concepts of distributed tracing, this is the book for you. If you are tasked to observe your .NET application, this book is a good start. And if you are working on implementing and tweaking telemetry, you will not be lost and this book will guide you.

Distributed tracing is a powerful tool for understanding how your applications work. It can help you identify performance bottlenecks, troubleshoot errors, and improve the overall reliability of your system. Historically, distributed tracing was just an add-on, an afterthought. There are a slew of great tools that you can use to enable it. But as any add-on, it was often a big project for any decent-size application. With the rise of microservices and cloud environments, and with the increased pace of frameworks development, many solutions started lagging behind. This is when it became clear that distributed tracing must be a common language, spoken natively by all frameworks, clouds, and apps. You can find all you need to know about distributed tracing and its common language in the first chapter of this book.

There are many languages in the world. What makes .NET stand out is that beyond the extensibility points allowing for a great ecosystem of libraries and tools, there are many carefully integrated and well-supported built-in primitives. This way, most of the needs of any app are covered, and developers can concentrate on business logic. Distributed tracing became this built-in component. Liudmila herself designed primitives and integrated those primitives with .NET. So she knows what she is writing about in *Chapters 2* and *3* of the book, showing how easy it is to get started with .NET applications observability.

I also enjoyed the often overlooked aspect of instrumenting the brownfield applications. Liudmila knows how hard the change is, especially in the world of .NET applications where the backward compatibility standards are so high. This is why every .NET developer will appreciate *Chapter 15* of the book.

Whether you're an architect and a seasoned developer, or just getting started with distributed tracing, this book is an essential resource. I highly recommend it to anyone who wants to improve the performance and reliability of their .NET applications.

Sincerely,

Sergey Kanzhelev

Co-founder of OpenTelemetry and Co-chair of the W3C Distributed Tracing Working Group

Contributors

About the author

Liudmila Molkova is a Principal Software Engineer at Microsoft working on observability and client libraries. She is a co-author of distributed tracing implementations across the .NET ecosystem including HTTP client, Azure Functions, and Application Insights SDK. She currently works in the Azure SDK team on improving the developer experience and plays an Observability Architect role in the team. She's also an active contributor to OpenTelemetry semantic conventions and instrumentation working groups.

Liudmila's love of observability started at Skype, where she got first-hand experience in running complex systems at a high scale and was fascinated by how much telemetry can reveal even to those deeply familiar with the code.

I'm deeply grateful to those who made this book possible. I would like to thank Sergey for his mentorship and support; Vance and David for trailblazing the distributed tracing on .NET; Noah, Sourabh, Jamie, and Joy for their insightful feedback; Irina for making me believe in myself; and the amazing Packt team for their support and encouragement along the way. Most of all, my husband Pavel, as his reassurance and care were indispensable. Thank you!

About the reviewers

Joy Rathnayake is a Solutions Architect with over 20 years' experience and a part of the Solution Architecture team in WSO2 Inc., Colombo. He has experience in architecting, designing, and developing software solutions using Microsoft and related technologies. Joy has a professional diploma in software engineering from NIIT.

He is a recognized Microsoft **Most Valuable Professional** (**MVP**) and **Microsoft Certified Trainer** (**MCT**). He has contributed to developing content for Microsoft Certifications and worked as an SME for several Microsoft exam development projects. He is a passionate speaker and has presented at various events.

In his spare time, Joy enjoys writing blogs, making videos, and reading. You can connect with him on LinkedIn.

Jamie Taylor (@podcasterJay) is an accomplished software developer, esteemed host of *The .NET Core Podcast*, a Microsoft MVP, and the proud recipient of the prestigious Managing Director of the Year award for 2023. With over a decade of experience in the industry, Jamie has garnered a reputation as a skilled and visionary leader in the field of modern .NET development.

Jamie's expertise in software engineering, cross-platform development, and cloud hosting has positioned him as a sought-after speaker and thought leader in the .NET community. Through his podcast, he shares his wealth of knowledge, engaging listeners with his ability to simplify complex concepts and provide practical insights.

Sourabh Shirhatti is a product manager specializing in developer tools and frameworks. He currently works on the Developer Platform Team at Uber, building compelling experiences for Uber developers. Previously, he worked in the Developer Division at Microsoft, where he actively worked on observability features, including OpenTelemetry support for .NET. Sourabh's passion lies in creating intuitive and efficient experiences for developers, enabling them to build high-quality software. Outside of work, he enjoys exploring Seattle's vibrant culture and beautiful outdoors with his wife.

Table of Contents

3

The .NET Observability Ecosystem   49

4

Low-Level Performance Analysis with Diagnostic Tools   69

Part 2: Instrumenting .NET Applications

5

Configuration and Control Plane   91

Part 3: Observability for Common Cloud Scenarios

9

Best Practices   171

10

Tracing Network Calls   181

11

Instrumenting Messaging Scenarios   201

12

Instrumenting Database Calls   225

Part 4: Implementing Distributed Tracing in Your Organization

13

Driving Change   247

14

Creating Your Own Conventions 259

15

Instrumenting Brownfield Applications 273

Preface

If you have worked on distributed applications, infrastructure, or client libraries, you've likely encountered numerous ways in which distributed systems can break.

For example, a default retry policy on your service can bring it down along with all its dependencies. Race conditions can lead to deadlocks under certain load or result in data leak between user accounts. User operations that usually take milliseconds can significantly slow down, while service dashboards show no signs of other issues. Functional problems can cause obscure and inexplicable effects on the user's end.

When working with distributed applications, we rely on telemetry to assess their performance and functionality. We need even more telemetry to identify and mitigate issues.

In the past, we relied on custom logs and metrics collected using vendor-specific SDKs. We built custom parsers, processing pipelines, and reporting tools to make telemetry usable.

However, as applications have become more complex, we require better and more user-friendly approaches to understand what is happening in our systems. Personally, I find it unproductive to read through megabytes of logs or visually detect anomalies in metrics.

Distributed tracing is a technique that allows us to trace operations throughout the entire system. It provides correlation and causation to our telemetry, enabling us to retrieve all the relevant data describing a specific operation or find all operations based on the context, such as a requested resource or a user identifier.

Distributed tracing alone is not enough; we need other telemetry signals such as metrics, events, logs, and profiles, as well as libraries to collect and export them to observability backends. Fortunately, we have OpenTelemetry for this purpose. OpenTelemetry is a cloud-native, vendor-neutral telemetry platform available in multiple programming languages. It offers the core components necessary to collect custom data along with instrumentation libraries for common technologies. OpenTelemetry standardizes telemetry formats for different signals ensuring correlation, consistency, and structure in the collected data.

By leveraging consistent and structured telemetry, different observability vendors can provide tools such as service maps, trace visualizations, error classification, and detection of common properties contributing to failures. This essentially allows us to automate the error-prone and tedious parts of performance analysis that humans struggle with. Monitoring and debugging techniques can now become standardized practices across the industry, no longer relying on tribal knowledge, runbooks, or outdated documentation.

Modern Distributed Tracing in .NET explores all aspects of telemetry collection in .NET applications, with a focus on distributed tracing and performance analysis. It begins with an overview of the observability challenges and solutions and then delves into the built-in monitoring capabilities offered by modern .NET applications. These capabilities become even more impressive when used alongside OpenTelemetry. While shared OpenTelemetry instrumentation libraries can take us a long way, sometimes we still need to write custom instrumentations. The book shows how to collect custom traces, metrics, and logs while considering performance impact and verbosity. It also covers the instrumentation of common cloud patterns such as network calls, messaging, and database interactions. Finally, it discusses the organizational and technical aspects of implementing and evolving observability in existing systems.

The observability field is still relatively new and rapidly evolving, which means there are often multiple solutions available for almost any problem. This book aims to explain fundamental observability concepts and provides several possible solutions to common problems while highlighting the associated trade-offs. It also helps you gain practical skills to implement and leverage tracing and observability.

I hope you find the provided examples useful and use them as a playground for experimentation. I encourage you to explore new and creative approaches to making distributed systems more observable and to share your findings with the community!

Who this book is for

This book is for software developers, architects, and system operators running .NET services who want to use modern observability tools and standards. It offers a holistic approach to performance analysis and end-to-end debugging. Software testers and support engineers will also find this book useful. Basic knowledge of the C# programming language and the .NET platform is assumed for grasping the examples of manual instrumentation, but is not necessary.

What this book covers

Chapter 1, Observability Needs of Modern Applications, provides an overview of common monitoring techniques and introduces distributed tracing. It covers OpenTelemetry – a vendor-agnostic telemetry platform and shows how it addresses observability challenges of distributed applications with correlated telemetry signals.

Chapter 2, Native Monitoring in .NET, offers an overview of the diagnostic capabilities provided by .NET out-of-the-box. These capabilities include structured and correlated logs and counters along with ad-hoc monitoring with the dotnet-monitor tool. We'll also instrument the first application with OpenTelemetry and get hands-on experience with distributed tracing.

Chapter 3, The .NET Observability Ecosystem, explores a broader set of tracing instrumentations and environments. We'll learn how to find and evaluate instrumentation libraries, get traces from infrastructure such as Dapr, and finally instrument serverless applications using AWS Lambda and Azure Functions as examples.

Chapter 4, Low-Level Performance Analysis with Diagnostic Tools, provides an introduction into lower-level .NET diagnostics and performance analysis. We'll see how to collect and analyze runtime counters and performance traces to get more observability within the process when distributed tracing does not provide enough input.

Chapter 5, Configuration and Control Plane, provides an overview of OpenTelemetry configuration and customization. We'll explore different sampling strategies and learn how to enrich and filter spans or customize metrics collection. Finally, we'll introduce OpenTelemetry Collector – an agent that can take care of many telemetry post-processing tasks.

Chapter 6, Tracing Your Code, dives into tracing instrumentation with .NET tracing APIs or OpenTelemetry shim. Here, we'll learn about the `Activity` and `ActivitySource` classes used to collect spans, show how to leverage ambient context propagation within the process, and record events and exceptions. We'll also cover integration testing for your instrumentation code.

Chapter 7, Adding Custom Metrics, delves into the modern .NET metrics API. You'll learn about available instruments - counters, gauges, and histograms used to aggregate measurements in different ways and get hands-on experience implementing and using metrics to monitor system health or to investigate performance issues.

Chapter 8, Writing Structured and Correlated Logs, provides an overview of logging in .NET focusing on `Microsoft.Extension.Logging`. We'll learn to write structured and queryable logs efficiently and collect them with OpenTelemetry. We'll also look into managing logging costs using OpenTelemetry Collector.

Chapter 9, Best Practices, provides guidance on choosing most suitable telemetry signals depending on application needs and scenarios, and shows how to control telemetry costs with minimal impact on observability. It also introduces OpenTelemetry semantic conventions – telemetry collection recipes for common patterns and technologies.

Chapter 10, Tracing Network Calls, explores network call instrumentation using gRPC as an example. We'll learn how to instrument simple request-response calls following RPC semantic conventions and propagate context. We'll also cover challenges and possible solutions when instrumenting streaming calls.

Chapter 11, Instrumenting Messaging Scenarios, explores instrumentation for asynchronous processing scenarios. We'll learn how to trace messages end-to-end, instrument batching scenarios, and introduce messaging-specific metrics allowing to detect scaling and performance issues.

Chapter 12, Instrumenting Database Calls, explores database and cache instrumentation with tracing and metrics. We'll also cover forwarding external metrics from a Redis instance into our observability backend and use the collected telemetry for performance analysis and caching strategy optimization.

Chapter 13, Driving Change, covers organizational and planning aspects related to observability improvements. We'll discuss the cost of low observability and suggest several ways to measure them. We'll come up with an onboarding plan, talk about common pitfalls, and see how to benefit from better observability in daily development tasks.

Chapter 14, Creating Your Own Conventions, provides suggestions on how to collect telemetry consistently across the system starting with a unified OpenTelemetry configuration. We'll also learn to define custom semantic conventions and implement them in shared code, making it easy to follow them.

Chapter 15, Instrumenting Brownfield Applications, discusses challenges with instrumenting newer part of the system in presence of legacy services. We'll suggest solutions that can minimize changes to legacy components and learn to leverage legacy correlation propagation formats, implement minimalistic pass-through context propagation, and forward telemetry from legacy services to the new backend.

To get the most out of this book

The examples in this book were developed with .NET 7.0. Most of them are cross-platform and run in Docker Linux containers or with the `dotnet` CLI tools. Examples were tested on Windows OS with OpenTelemetry version 1.4.0. They should work with future versions of .NET and OpenTelemetry libraries. We use pinned versions of OpenTelemetry Collector, Prometheus, Jaeger, and other external images in the `docker-compose` files.

In *Chapter 3, The .NET Observability Ecosystem*, we will experiment with AWS and Azure client libraries and serverless environments. AWS and/or Azure subscriptions are recommended, but not essential. We will stay within the free tier on AWS and within promotional credits amount available on Azure.

Software/hardware covered in the book	Operating system requirements
.NET SDK 7.0	Windows, macOS, or Linux
OpenTelemetry for .NET version 1.4.0	Windows, macOS, or Linux
Docker and `docker-compose` tools	Windows, macOS, or Linux
.NET Framework 4.6.2 (used in an example of a legacy system in *Chapter 15*)	Windows
PerfView tool (in *Chapter 4*)	Windows, cross-platform alternatives are available

While OpenTelemetry guarantees API compatibility in future versions, the semantic conventions mentioned in this book are not stable. So spans, metrics, events, and attributes may be renamed or changed in a different way. Please refer to OpenTelemetry specification repo (`https://github. com/open-telemetry/opentelemetry-specification`) to find what's new in the semantic conventions area.

If you are using the digital version of this book, we advise you to type the code yourself or access the code from the book's GitHub repository (a link is available in the next section). Doing so will help you avoid any potential errors related to the copying and pasting of code.

Download the example code files

You can download the example code files for this book from GitHub at `https://github.com/PacktPublishing/Modern-Distributed-Tracing-in-.NET`. If there's an update to the code, it will be updated in the GitHub repository.

We also have other code bundles from our rich catalog of books and videos available at `https://github.com/PacktPublishing/`. Check them out!

Code in Action

The Code in Action videos for this book can be viewed at `https://packt.link/O10rj`.

Download the color images

We also provide a PDF file that has color images of the screenshots and diagrams used in this book. You can download it here: `https://packt.link/BBBNm`.

Conventions used

There are a number of text conventions used throughout this book.

`Code in text`: Indicates code words in text, database table names, folder names, filenames, file extensions, pathnames, dummy URLs, user input, and Twitter handles. Here is an example: "Another option is to pass the `traceparent` value in W3C Trace Context format to the `StartActivity` method as a string."

A block of code is set as follows:

```
using var activity = Source.StartActivity("DoWork");
try
{
  await DoWorkImpl(workItemId);
}
catch
{
  activity?.SetStatus(ActivityStatusCode.Error);
}
```

When we wish to draw your attention to a particular part of a code block, the relevant lines or items are set in bold:

```
using var provider = Sdk.CreateTracerProviderBuilder()
  .ConfigureResource(b => b.AddService("sample"))
  .AddSource("Worker")
  .AddJaegerExporter()
```

Any command-line input or output is written as follows:

```
$ docker-compose up --build
$ dotnet run
```

Bold: Indicates a new term, an important word, or words that you see onscreen. For instance, words in menus or dialog boxes appear in **bold**. Here is an example: "Let's open the trace file with PerfView and then click on the **Thread Time** option."

> **Tips or important notes**
> Appear like this.

Get in touch

Feedback from our readers is always welcome.

General feedback: If you have questions about any aspect of this book, email us at customercare@packtpub.com and mention the book title in the subject of your message.

Errata: Although we have taken every care to ensure the accuracy of our content, mistakes do happen. If you have found a mistake in this book, we would be grateful if you would report this to us. Please visit www.packtpub.com/support/errata and fill in the form.

Piracy: If you come across any illegal copies of our works in any form on the internet, we would be grateful if you would provide us with the location address or website name. Please contact us at copyright@packt.com with a link to the material.

If you are interested in becoming an author: If there is a topic that you have expertise in and you are interested in either writing or contributing to a book, please visit authors.packtpub.com.

Share Your Thoughts

Once you've read *Modern Distributed Tracing in .NET*, we'd love to hear your thoughts! Scan the QR code below to go straight to the Amazon review page for this book and share your feedback.

https://packt.link/r/1-837-63613-3

Your review is important to us and the tech community and will help us make sure we're delivering excellent quality content.

Download a free PDF copy of this book

Thanks for purchasing this book!

Do you like to read on the go but are unable to carry your print books everywhere?

Is your eBook purchase not compatible with the device of your choice?

Don't worry, now with every Packt book you get a DRM-free PDF version of that book at no cost.

Read anywhere, any place, on any device. Search, copy, and paste code from your favorite technical books directly into your application.

The perks don't stop there, you can get exclusive access to discounts, newsletters, and great free content in your inbox daily

Follow these simple steps to get the benefits:

1. Scan the QR code or visit the link below

https://packt.link/free-ebook/9781837636136

2. Submit your proof of purchase
3. That's it! We'll send your free PDF and other benefits to your email directly

Part 1: Introducing Distributed Tracing

In this part, we'll introduce the core concepts of distributed tracing and demonstrate how it makes running cloud applications easier. We'll auto-instrument our first service and explore the .NET approach to observability, built around OpenTelemetry.

This part has the following chapters:

1
Observability Needs of Modern Applications

With the increasing complexity of distributed systems, we need better tools to build and operate our applications. **Distributed tracing** is one such technique that allows you to collect structured and correlated telemetry with minimum effort and enables observability vendors to build powerful analytics and automation.

In this chapter, we'll explore common observability challenges and see how distributed tracing brings observability to our systems where logs and counters can't. We'll see how correlation and causation along with structured and consistent telemetry help answer arbitrary questions about the system and mitigate issues faster.

Here's what you will learn:

- An overview of monitoring techniques using counters, logs, and events
- Core concepts of distributed tracing – the span and its structure
- Context propagation standards
- How to generate meaningful and consistent telemetry
- How to use distributed tracing along with metrics and logs for performance analysis and debugging

By the end of this chapter, you will become familiar with the core concepts and building blocks of distributed tracing, which you will be able to use along with other telemetry signals to debug functional issues and investigate performance issues in distributed applications.

Understanding why logs and counters are not enough

Monitoring and observability cultures vary across the industry; some teams use ad hoc debugging with `printf` while others employ sophisticated observability solutions and automation. Still, almost every system uses a combination of common telemetry signals: logs, events, metrics or counters, and profiles. Telemetry collection alone is not enough. A system is **observable** if we can detect and investigate issues, and to achieve this, we need tools to store, index, visualize, and query the telemetry, navigate across different signals, and automate repetitive analysis.

Before we begin exploring tracing and discovering how it helps, let's talk about other telemetry signals and their limitations.

Logs

A **log** is a record of some event. Logs typically have a timestamp, level, class name, and formatted message, and may also have a property bag with additional context.

Logs are a low-ceremony tool, with plenty of logging libraries and tools for any ecosystem.

Common problems with logging include the following:

- **Verbosity**: Initially, we won't have enough logs, but eventually, as we fill gaps, we will have too many. They become hard to read and expensive to store.

- **Performance**: Logging is a common performance issue even when used wisely. It's also very common to serialize objects or allocate strings for logging even when the logging level is disabled.

   One new log statement can take your production down; I did it once. The log I added was written every millisecond. Multiplied by a number of service instances, it created an I/O bottleneck big enough to significantly increase latency and the error rate for users.

- **Not queryable**: Logs coming from applications are intended for humans. We can add context and unify the format within our application and still only be able to filter logs by context properties. Logs change with every refactoring, disappear, or become out of date. New people joining a team need to learn logging semantics specific to a system, and the learning curve can be steep.

- **No correlation**: Logs for different operations are interleaved. The process of finding logs describing certain operations is called correlation. In general, log correlation, especially across services, must be implemented manually (spoiler: not in ASP.NET Core).

> **Note**
> Logs are easy to produce but are verbose, and then can significantly impact performance. They are also difficult to filter, query, or visualize.

To be accessible and useful, logs are sent to some central place, a **log management system**, which stores, parses, and indexes them so they can be queried. This implies that your logs need to have at least some structure.

`ILogger` in .NET supports structured logging, as we'll see in *Chapter 8, Writing Structured and Correlated Logs*, so you get the human-readable message, along with the context. Structured logging, combined with structured storage and indexing, converts your logs into rich events that you can use for almost anything.

Events

An **event** is a structured record of something. It has a timestamp and a property bag. It may have a name, or that could just be one of the properties.

The difference between logs and events is semantical – an event is structured and usually follows a specific schema.

For example, an event that describes adding an item to a shopping bag should have a well-known name, such as `shopping_bag_add_item` with `user-id` and `item-id` properties. Then, you can query them by name, item, and user. For example, you can find the top 10 popular items across all users.

If you write it as a log message, you'd probably write something like this:

```
logger.LogInformation("Added '{item-id}' to shopping bag
    for '{user-id}'", itemId, userId)
```

If your logging provider captures individual properties, you would get the same context as with events. So, now we can find every log for this user and item, which probably includes other logs not related to adding an item.

> **Note**
> Events with consistent schema can be queried efficiently but have the same verbosity and performance problems as logs.

Metrics and counters

Logs and events share the same problem – verbosity and performance overhead. One way to solve them is aggregation.

A **metric** is a value of something aggregated by dimensions and over a period of time. For example, a request latency metric can have an HTTP route, status code, method, service name, and instance dimensions.

Common problems with metrics include the following:

- **Cardinality**: Each combination of dimensions is a time series, and aggregation happens within one time series. Adding a new dimension causes a combinatorial explosion, so metrics must have low cardinality – that is, they cannot have too many dimensions, and each one must have a small number of distinct values. As a result, you can't measure granular things such as per-user experience with metrics.

- **No causation**: Metrics only show correlation and no cause and effect, so they are not a great tool to investigate issues.

  As an expert on your system, you might use your intuition to come up with possible reasons for certain types of behavior and then use metrics to confirm your hypothesis.

- **Verbosity**: Metrics have problems with verbosity too. It's common to add metrics that measure just one thing, such as `queue_is_full` or `queue_is_empty`. Something such as `queue_utilization` would be more generic. Over time, the number of metrics grows along with the number of alerts, dashboards, and team processes relying on them.

> **Note**
> Metrics have low impact on performance, low volume that doesn't grow much with scale, low storage costs, and low query time. They are great for dashboards and alerts but not for issue investigation or granular analytics.

A **counter** is a single time series – it's a metric without dimensions, typically used to collect resource utilization such as CPU load or memory usage. Counters don't work well for application performance or usage, as you need a dedicated counter per each combination of attributes, such as HTTP route, status code, and method. It is difficult to collect and even harder to use. Luckily, .NET supports metrics with dimensions, and we will discuss them in *Chapter 7, Adding Custom Metrics*.

What's missing?

Now you know all you need to monitor a monolith or small distributed system – use metrics for system health analysis and alerts, events for usage, and logs for debugging. This approach has taken the tech industry far, and there is nothing essentially wrong with it.

With up-to-date documentation, a few key performance and usage metrics, concise, structured, correlated, and consistent events, common conventions, and tools across all services, anyone operating your system can do performance analysis and debug issues.

> **Note**
>
> *So, the ultimate goal is to efficiently operate a system, and the problem is not a specific telemetry signal or its limitations but a lack of standard solutions and practices, correlation, and structure for existing signals.*

Before we jump into distributed tracing and see how its ecosystem addresses these gaps, let's summarize the new requirements we have for the perfect observability solution we intend to solve with tracing and the new capabilities it brings. Also, we should keep in mind the old capabilities – low-performance overhead and manageable costs.

Systematic debugging

We need to be able to investigate issues in a generic way. From an error report to an alert on a metric, we should be able to drill down into the issue, follow specific requests end to end, or bubble up from an error deep in the stack to understand its effect on users.

All this should be reasonably easy to do when you're on call and paged at 2AM to resolve an incident in production.

Answering ad hoc questions

I might want to understand whether users from Redmond, WA, who purchased a product from my website are experiencing longer delivery times than usual and why – because of the shipment company, rain, cloud provider issues in this region, or anything else.

It should not be required to add more telemetry to answer most of the usage or performance questions. Occasionally, you'd need to add a new context property or an event, but it should be rare on a stable code path.

Self-documenting systems

Modern systems are dynamic – with continuous deployments, feature flag changes in runtime, and dozens of external dependencies with their own instabilities, nobody can know everything.

Telemetry becomes your single source of truth. Assuming it has enough context and common semantics, an observability vendor should be able to visualize it reasonably well.

Auto-instrumentation

It's difficult to instrument everything in your system – it's repetitive, error-prone, and hard to keep up to date, test, and enforce common schema and semantics. We need shared instrumentations for common libraries, while we would only add application-specific telemetry and context.

With an understanding of these requirements, we will move on to distributed tracing.

Introducing distributed tracing

Distributed tracing is a technique that brings structure, correlation and causation to collected telemetry. It defines a special event called *span* and specifies causal relationships between spans. Spans follow common conventions that are used to visualize and analyze traces.

Span

A **span** describes an operation such as an incoming or outgoing HTTP request, a database call, an expensive I/O call, or any other interesting call. It has just enough structure to represent anything and still be useful. Here are the most important span properties:

- The span's name should describe the operation type in human-readable format, have low cardinality, and be human-readable.

- The span's start time and duration.

- The status indicates success, failure, or no status.

- The span kind distinguishes the client, server, and internal calls, or the producer and consumer for async scenarios.

- Attributes (also known as tags or annotations) describe specific operations.

- Span context identifies spans and is propagated everywhere, enabling correlation. A parent span identifier is also included on child spans for causation.

- Events provide additional information about operations within a span.

- Links connect traces and spans when parent-child relationships don't work – for example, for batching scenarios.

> **Note**
> In .NET, the tracing span is represented by `System.Diagnostics.Activity`. The `System.Span` class is not related to distributed tracing.

Relationships between spans

A span is a unit of tracing, and to trace more complex operations, we need multiple spans.

For example, a user may attempt to get an image and send a request to the service. The image is not cached, and the service requests it from the cold storage (as shown in *Figure 1.1*):

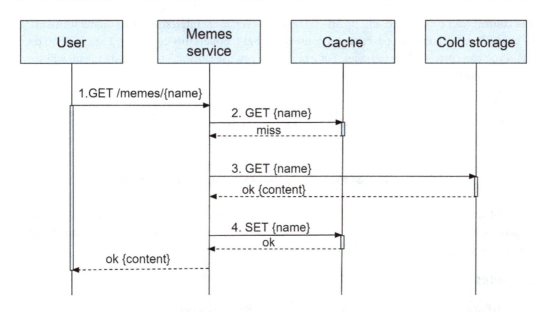

Figure 1.1 – A GET image request flow

To make this operation debuggable, we should report multiple spans:

1. The incoming request

2. The attempt to get the image from the cache

3. Image retrieval from the cold storage

4. Caching the image

These spans form a **trace** – a set of related spans fully describing a logical end-to-end operation sharing the same `trace-id`. Within the trace, each span is identified by `span-id`. Spans include a pointer to a parent span – it's just their parent's `span-id`.

`trace-id`, `span-id`, and `parent-span-id` allow us to not only correlate spans but also record relationships between them. For example, in *Figure 1.2*, we can see that Redis GET, SETEX, and HTTP GET spans are siblings and the incoming request is their parent:

Figure 1.2 – Trace visualization showing relationships between spans

Spans can have more complicated relationships, which we'll talk about later in *Chapter 6, Tracing Your Code*.

Span context (aka `trace-id` and `span-id`) enables even more interesting cross-signal scenarios. For example, you can stamp parent span context on logs (spoiler: just configure `ILogger` to do it) and you can correlate logs to traces. For example, if you use `ConsoleProvider`, you will see something like this:

```
info: storage.Controllers.MemesController[0]
      => SpanId:71a034fd851470ac, TraceId:720ac9dabcef7ba1827efd3e69882e44 => ConnectionId:
0HMMCU56FJOU0 => RequestPath:/memes/dotnet RequestId:0HMMCU56FJOU0:00000002 => storage.Cont
rollers.MemesController.Get (storage)
      Returning 'dotnet', 1516 bytes
```

Figure 1.3 – Logs include span context and can be correlated to other signals

You could also link metrics to traces using exemplars – metric metadata containing the trace context of operations that contributed to a recorded measurement. For instance, you can check examples of spans that correspond to the long tail of your latency distribution.

Attributes

Span attributes are a property bag that contains details about the operation.

Span attributes should describe this specific operation well enough to understand what happened. OpenTelemetry semantic conventions specify attributes for popular technologies to help with this, which we'll talk about in the *Ensuring consistency and structure* section later in this chapter.

For example, an incoming HTTP request is identified with at least the following attributes: the HTTP method, path, query, API route, and status code:

Memes/{name}   Service: **storage**   Duration: **117.86ms**   Start Time: **16.94ms**

∨ **Tags**

http.flavor	1.1
http.host	storage:5051
http.method	GET
http.route	Memes/{name}
http.scheme	http
http.status_code	200
http.target	/memes/4d7aacc88105
http.url	http://storage:5051/memes/4d7aacc88105
internal.span.format	proto
otel.library.name	OpenTelemetry.Instrumentation.AspNetCore
otel.library.version	1.0.0.0
span.kind	server

Figure 1.4 – The HTTP server span attributes

Instrumentation points

So, we have defined a span and its properties, but when should we create spans? Which attributes should we put on them? While there is no strict standard to follow, here's the rule of thumb:

Create a new span for every incoming and outgoing network call and use standard attributes for the protocol or technology whenever available.

This is what we've done previously with the memes example, and it allows us to see what happened on the service boundaries and detect common problems: dependency issues, status, latency, and errors on each service. This also allows us to correlate logs, events, and anything else we collect. Plus, observability backends are aware of HTTP semantics and will know how to interpret and visualize your spans.

There are exceptions to this rule, such as socket calls, where requests could be too small to be instrumented. In other cases, you might still be rightfully concerned with verbosity and the volume of generated data – we'll see how to control it with sampling in *Chapter 5, Configuration and Control Plane*.

Tracing – building blocks

Now that you are familiar with the core concepts of tracing and its methodology, let's talk about implementation. We need a set of convenient APIs to create and enrich spans and pass context around. Historically, every **Application Performance Monitoring** (**APM**) tool had its own SDKs to collect telemetry with their own APIs. Changing the APM vendor meant rewriting all your instrumentation code.

OpenTelemetry solves this problem – it's a cross-language telemetry platform for tracing, metrics, events, and logs that unifies telemetry collection. Most of the APM tools, log management, and observability backends support OpenTelemetry, so you can change vendors without rewriting any instrumentation code.

.NET tracing implementation conforms to the OpenTelemetry API specification, and in this book, .NET tracing APIs and OpenTelemetry APIs are used interchangeably. We'll talk about the difference between them in *Chapter 6, Tracing Your Code*.

Even though OpenTelemetry primitives are baked into .NET and the instrumentation code does not depend on them, to collect telemetry from the application, we still need to add the **OpenTelemetry SDK**, which has everything we need to configure a collection and an exporter. You might as well write your own solution compatible with .NET tracing APIs.

OpenTelemetry became an industry standard for tracing and beyond; it's available in multiple languages, and in addition to a unified collection of APIs it provides configurable SDKs and a standard wire format for the telemetry – **OpenTelemetry protocol** (**OTLP**). You can send telemetry to any compatible vendor, either by adding a specific exporter or, if the backend supports OTLP, by configuring the vendor's endpoint.

As shown in *Figure 1.5*, the application configures the OpenTelemetry SDK to export telemetry to the observability backend. Application code, .NET libraries, and various instrumentations use .NET tracing APIs to create spans, which the OpenTelemetry SDK listens to, processes, and forwards to an exporter.

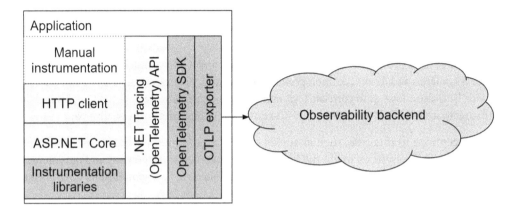

Figure 1.5 – Tracing building blocks

So, OpenTelemetry decouples instrumentation code from the observability vendor, but it does much more than that. Now, different applications can share instrumentation libraries and observability vendors have unified and structured telemetry on top of which they can build rich experiences.

Instrumentation

Historically, all APM vendors had to instrument popular libraries: HTTP clients, web frameworks, Entity Framework, SQL clients, Redis client libraries, RabbitMQ, cloud providers' SDKs, and so on. That did not scale well. But with .NET tracing APIs and OpenTelemetry semantics, instrumentation became common for all vendors. You can find a growing list of shared community instrumentations in the OpenTelemetry Contrib repo: https://github.com/open-telemetry/opentelemetry-dotnet-contrib.

Moreover, since OpenTelemetry is a vendor-neutral standard and baked into .NET, it's now possible for libraries to implement native instrumentation – HTTP and gRPC clients, ASP.NET Core, and several other libraries support it.

Even with native tracing support, it's off by default – you need to install and register specific instrumentation (which we'll cover in *Chapter 2*, *Native Monitoring in .NET*). Otherwise, tracing code does nothing and, thus, does not add any performance overhead.

Backends

The **observability backend** (aka monitoring, APM tool, and log management system) is a set of tools responsible for ingestion, storage, indexing, visualization, querying, and probably other things that help you monitor your system, investigate issues, and analyze performance.

Observability vendors build these tools and provide rich user experiences to help you use traces along with other signals.

Collecting traces for common libraries became easy with the OpenTelemetry ecosystem. As you'll see in *Chapter 2, Native Monitoring in .NET*, most of it can be done automatically with just a few lines of code at startup. But how do we use them?

While you can send spans to `stdout` and store them on the filesystem, this would not leverage all tracing benefits. Traces can be huge, but even when they are small, grepping them is not convenient.

Tracing visualizations (such as a Gantt chart, trace viewer, or trace timeline) is one of the common features tracing providers have. *Figure 1.6* shows a trace timeline in Jaeger – an open source distributed tracing platform:

Figure 1.6 – Trace visualization in Jaeger with errors marked with exclamation point

While it may take a while to find an error log, the visualization shows what's important – where failures are, latency, and a sequence of steps. As we can see in *Figure 1.6*, the frontend call failed because of failure on the storage side, which we can further drill into.

However, we can also see that the frontend made four consecutive calls into storage, which potentially could be done in parallel to speed things up.

Another common feature is filtering or querying by any of the span properties such as name, `trace-id`, `span-id`, `parent-id`, name, attribute name, status, timestamp, duration, or anything else. An example of such a query is shown in *Figure 1.7*:

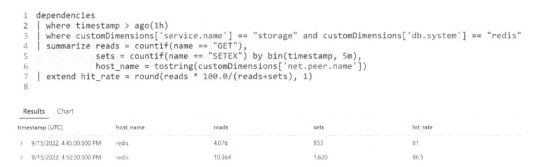

Figure 1.7 – A custom Azure Monitor query that calculates the Redis hit rate

For example, we don't report a metric for the cache hit rate, but we can estimate it from traces. While they're not precise because of sampling and might be more expensive to query than metrics, we can still do it ad hoc, especially when we investigate specific failures.

Since traces, metrics, and logs are correlated, you will fully leverage observability capabilities if your vendor supports multiple signals or integrates well with other tools.

Reviewing context propagation

Correlation and causation are the foundation of distributed tracing. We've just covered how related spans share the same `trace-id` and have a pointer to the parent recorded in `parent-span-id`, forming a casual chain of operations. Now, let's explore how it works in practice.

In-process propagation

Even within a single service, we usually have nested spans. For example, if we trace a request to a REST service that just reads an item from a database, we'd want to see at least two spans – one for an incoming HTTP request and another for a database query. To correlate them properly, we need to pass span context from ASP.NET Core to the database driver.

One option is to pass context explicitly as a function argument. It's a viable solution in Go, where explicit context propagation is a standard, but in .NET, it would make onboarding onto distributed tracing difficult and would ruin the auto-instrumentation magic.

.NET Activity (aka the span) is propagated implicitly. Current activity can always be accessed via the `Activity.Current` property, backed up by `System.Threading.AsyncLocal<T>`.

Using our previous example of a service reading from the database, ASP.NET Core creates an Activity for the incoming request, and it becomes current for anything that happens within the scope of this request. Instrumentation for the database driver creates another one that uses `Activity.Current` as its parent, without knowing anything about ASP.NET Core and without the user application passing the Activity around. The logging framework would stamp `trace-id` and `span-id` from `Activity.Current`, if configured to do so.

It works for sync or async code, but if you process items in the background using in-memory queues or manipulate with threads explicitly, you would have to help runtime and propagate activities explicitly. We'll talk more about it in *Chapter 6, Tracing Your Code*.

Out-of-process propagation

In-process correlation is awesome, and for monolith applications, it would be almost sufficient. But in the microservice world, we need to trace requests end to end and, therefore, propagate context over the wire, and here's where standards come into play.

You can find multiple practices in this space – every complex system used to support something custom, such as `x-correlation-id` or `x-request-id`. You can find `x-cloud-trace-context` or `grpc-trace-bin` in old Google systems, `X-Amzn-Trace-Id` on AWS, and `Request-Id` variations and `ms-cv` in the Microsoft ecosystem. Assuming your system is heterogeneous and uses a variety of cloud providers and tracing tools, correlation is difficult.

Trace context (which you can explore in more detail at `https://www.w3.org/TR/trace-context`) is a relatively new standard, converting context propagation over HTTP, but it's widely adopted and used by default in OpenTelemetry and .NET.

W3C Trace Context

The trace context standard defines `traceparent` and `tracestate` HTTP headers and the format to populate context on them.

The traceparent header

The `traceparent` is an HTTP request header that carries the protocol version, `trace-id`, `parent-id`, and `trace-flags` in the following format:

```
traceparent: {version}-{trace-id}-{parent-id}-{trace-flags}
```

- `version`: The protocol version – only `00` is defined at the moment.
- `trace-id`: The logical end-to-end operation ID.
- `parent-id`: Identifies the client span and serves as a parent for the corresponding server span.

- `trace-flags`: Represents the sampling decision (which we'll talk about in *Chapter 5, Configuration and Control Plane*). For now, we can determine that `00` indicates that the parent span was sampled out and `01` means it was sampled in.

All identifiers must be present – that is, `traceparent` has a fixed length and is easy to parse. *Figure 1.8* shows an example of context propagation with the `traceparent` header:

Figure 1.8 – traceparent is populated from the outgoing span
context and becomes a parent for the incoming span

> **Note**
>
> The protocol does not require creating spans and does not specify instrumentation points. Common practice is to create spans per outgoing and incoming requests, and put client span context into request headers.

The tracestate header

The `tracestate` is another request header, which carries additional context for the tracing tool to use. It's *designed for OpenTelemetry or an APM tool to carry additional control information and not for application-specific context* (covered in detail later in the *Baggage* section).

The `tracestate` consists of a list of key-value pairs, serialized to a string with the following format: `"vendor1=value1,vendor2=value2"`.

The `tracestate` can be used to propagate incompatible legacy correlation IDs, or some additional identifiers vendor needs.

OpenTelemetry, for example, uses it to carry a sampling probability and score. For example, `tracestate: "ot=r:3;p:2"` represents a key-value pair, where the key is `ot` (OpenTelemetry tag) and the value is `r:3;p:2`.

The `tracestate` header has a soft limitation on size (512 characters) and can be truncated.

The traceresponse (draft) header

Unlike `traceparent` and `tracestate`, **traceresponse** is a response header. At the time of writing, it's defined in W3C Trace-Context Level 2 (`https://www.w3.org/TR/trace-context-2/`) and has reached W3C Editor's Draft status. There is no support for it in .NET or OpenTelemetry.

`traceresponse` is very similar to `traceparent`. It has the same format, but instead of client-side identifiers, it returns the `trace-id` and `span-id` values of the server span:

```
traceresponse: 00-{trace-id}-{child-id}-{trace-flags}
```

`traceresponse` is optional in the sense that the server does not need to return it, even if it supports W3C Trace-Context Level 2. It's useful to return `traceresponse` when the client did not pass a valid `traceparent`, but can log `traceresponse`.

External-facing services may decide to start a new trace, because they don't trust the caller's `trace-id` generation algorithm. Uniform random distribution is one concern; another reason could be a special `trace-id` format. If the service restarts a trace, it's a good idea to return the `traceresponse` header to caller.

B3

The B3 specification (`https://github.com/openzipkin/b3-propagation`) was adopted by Zipkin – one of the first distributed tracing systems.

B3 identifiers can be propagated as a single b3 header in the following format:

```
b3: {trace-id}-{span-id}-{sampling-state}-{parent-span-id}
```

Another way is to pass individual components, using `X-B3-TraceId`, `X-B3-SpanId`, `X-B3-ParentSpanId`, and `X-B3-Sampled`.

The sampling state suggests whether a service should trace the corresponding request. In addition to 0 (don't record) and 1 (do record), it allows us to force tracing with a flag set to d. It's usually done for debugging purposes. The sampling state can be passed without other identifiers to specify the desired sampling decision to the service.

> **Note**
>
> The key difference with W3C Trace-Context, beyond header names, is the presence of both `span-id` and `parent-span-id`. B3 systems can use the same `span-id` on the client and server sides, creating a single span for both.

Zipkin reuses `span-id` from the incoming request, also specifying `parent-span-id` on it. The Zipkin span represents the client and server at the same time, as shown in *Figure 1.9*, recording different durations and statuses for them:

Figure 1.9 – Zipkin creates one span to represent the client and server

OpenTelemetry and .NET support b3 headers but ignore `parent-span-id` – they generate a new `span-id` for every span, as it's not possible to reuse `span-id` (see *Figure 1.10*).

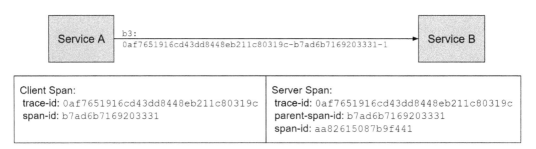

Figure 1.10 – OpenTelemetry does not use parent-span-id from B3
headers and creates different spans for the client and server

Baggage

So far, we have talked about span context and correlation. But in many cases, distributed systems have application-specific context. For example, you authorize users on your frontend service, and after that, `user-id` is not needed for application logic, but you still want to add it as an attribute on spans from all services to query and aggregate it on a per-user basis.

You can stamp `user-id` once on the frontend. Then, spans recorded on the backend will not have `user-id`, but they will share the same `trace-id` as the frontend. So, with some joins in your queries, you can still do per-user analysis. It works to some extent but may be expensive or slow, so you might decide to propagate `user-id` and stamp it on the backend spans too.

The **baggage** (`https://www.w3.org/TR/baggage/`) defines a generic propagation format for distributed context, and you can use it for business logic or anything else by adding, reading, removing, and modifying baggage members. For example, you can route requests to the test environment and pass feature flags or extra telemetry context.

Baggage consists of a list of semicolon-separated members. Each member has a key, value, and optional properties in the following formats – `key=value;property1;key2=property2` or `key=value;property1;key2=property2,anotherKey=anotherValue`.

OpenTelemetry and .NET only propagate baggage, but don't stamp it on any telemetry. You can configure `ILogger` to stamp baggage and need to enrich traces explicitly. We'll see how it works in *Chapter 5, Configuration and Control Plane*.

> **Tip**
>
> You should not put any sensitive information in baggage, as it's almost impossible to guarantee where it would flow – your application or sidecar infrastructure can forward it to your cloud provider or anywhere else.
>
> Maintain a list of well-known baggage keys across your system and only use known ones, as you might receive baggage from another system otherwise.

Baggage specification has a *working draft* status and may still change.

> **Note**
>
> While the W3C Trace Context standard is HTTP-specific and B3 applies to any RPC calls, they are commonly used for any context propagation needs – for example, they are passed as the event payload in messaging scenarios. This may change once protocol-specific standards are introduced.

Ensuring consistency and structure

As we already defined, spans are structured events describing interesting operations.

A span's start time, duration, status, kind, and context are strongly typed – they enable correlation and causation, allowing us to visualize traces and detect failures or latency issues.

The span's name and attributes describe an operation but are not strongly typed or strictly defined. If we don't populate them in a meaningful way, we can detect an issue but have no knowledge of what actually happened.

For example, for client HTTP calls, beyond generic properties, we want to capture at least the URL, method, and response code (or exception) – if we don't know any of these, we're blind. Once we populate them, we can start doing some powerful analysis with queries over such spans to answer the following common questions:

- Which dependency calls were made in the scope of this request? Which of them failed? What was the latency of each of them?

- Does my application make independent dependency calls in parallel or sequentially? Does it make any unnecessary requests when they can be done lazily?

- Are dependency endpoints configured correctly?

- What are the success or error rates and latency per dependency API?

> **Note**
>
> This analysis relies on an application using the same attributes for all HTTP dependencies. Otherwise, the operator that performs the queries will have a hard time writing and maintaining them.

With unified and community-driven telemetry collection taken off the observability vendor's plate, they can now fully focus on (semi-)automating analysis and giving us powerful performance and fault analysis tools.

OpenTelemetry defines a set of semantic conventions for spans, traces, and resources, which we'll talk more about in *Chapter 9*, *Best Practices*.

Building application topology

Distributed tracing, combined with semantic conventions, allows us to build visualizations such as an application map (aka service map), as shown in *Figure 1.11* – you could see your whole system along with key health metrics. It's an entry point to any investigation.

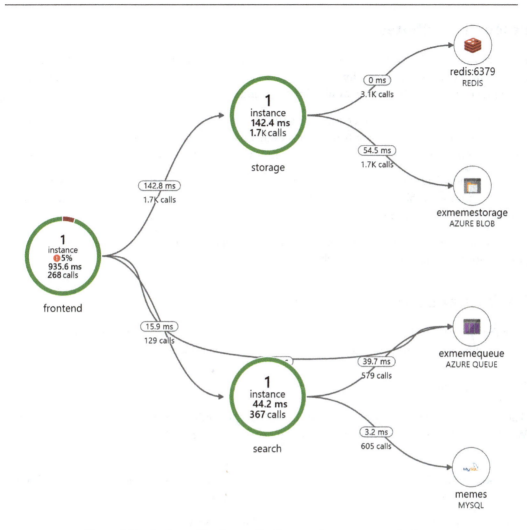

Figure 1.11 – An Azure Monitor application map for a meme service is an
up-to-date system diagram with all the basic health metrics

Observability vendors depend on trace and metrics semantics to build service maps. For example, the presence of HTTP attributes on the client span represents an outgoing HTTP call, and we need to show the outgoing arrow to a new dependency node. We should name this node based on the span's host attribute.

If we see the corresponding server span, we can now merge the server node with the dependency node, based on span context and causation. There are other visualizations or automation tools that you might find useful – for example, critical path analysis, or finding common attributes that correspond to higher latency or error rates. Each of these relies on span properties and attributes following common semantics or at least being consistent across services.

Resource attributes

Resource attributes describe the process, host, service, and environment, and are the same for all spans reported by the service instance – for example, the service name, version, unique service instance ID, cloud provider account ID, region, availability zone, and K8s metadata.

These attributes allow us to detect anomalies specific to certain environments or instances – for example, an error rate increase only on instances that have a new version of code, an instance that goes into a restart loop, or a cloud service in a region and availability zone that experiences issues.

Based on standard attributes, observability vendors can write generic queries to perform this analysis or build common dashboards. It also enables the community to create vendor-agnostic tools and solutions for popular technologies.

Such attributes describe a service instance and don't have to appear on every span – OTLP, for example, passes resource attributes once per batch of spans.

Performance analysis overview

Now that you know the core concepts around distributed tracing, let's see how we can use the observability stack to investigate common distributed system problems.

The baseline

Before we talk about problems, let's establish a baseline representing the behavior of a healthy system. We also need it to make data-driven decisions to help with common design and development tasks such as the following:

- **Risk estimation**: Any feature work on the hot path is a good candidate for additional performance testing prior to release and guarding new code with feature flags.

- **Capacity planning**: Knowing the current load is necessary to understand whether a system can handle planned growth and new features.

- **Understand improvement potential**: It makes more sense to optimize frequently executed code, as even small optimizations bring significant performance gains or cost reductions. Similarly, improving reliability brings the most benefits for components that have a higher error rate and that are used by other services.

- **Learning usage patterns**: Depending on how users interact with your system, you might change your scaling or caching strategy, extract specific functionality to a new service, or merge services.

Generic indicators that describe the performance of each service include the following:

- **Latency**: How fast a service responds
- **Throughput**: How many requests, events, or bytes the service is handling per second
- **Error rate**: How many errors a service returns

Your system might need other indicators to measure durability or data correctness.

Each of these signals is useful when it includes an API route, a status code, and other context properties. For example, the error rate could be low overall but high for specific users or API routes.

Measuring signals on the server and client sides, whenever possible, gives you a better picture. For example, you can detect network failures and avoid "*it works on my machine*" situations when clients see issues and servers don't.

Investigating performance issues

Let's divide performance issues into two overlapping categories:

- Widespread issues that affect a whole instance, server, or even the system, and move the distribution median.
- An individual request or job that takes too much time to complete. If we visualize the latency distribution, as shown in *Figure 1.12*, we'll see such issues in the long tail of distribution – they are rare, but part of normal behavior.

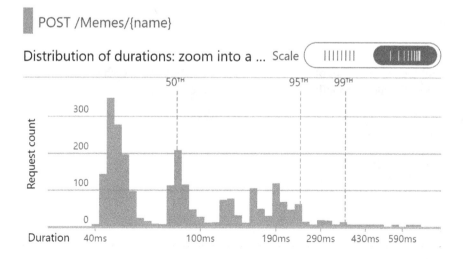

Figure 1.12 – Azure Monitor latency distribution visualization, with a median request
(the 50th percentile) taking around 80 ms and the 95th percentile around 250 ms

Long tails

Individual issues can be caused by an unfortunate chain of events – transient network issues, high contention in optimistic concurrency algorithms, hardware failures, and so on.

Distributed tracing is an excellent tool to investigate such issues. If you have a bug report, you might have a trace context for a problematic operation. To achieve it, make sure you show the `traceparent` value on the web page and return `traceresponse` or a document that users need to record, or log `traceresponse` when sending requests to your service.

So, if you know the trace context, you can start by checking the trace view. For example, in *Figure 1.13*, you can see an example of a long request caused by transient network issues.

Figure 1.13 – A request with high latency caused by transient network issues and retries

The frontend request took about 2.6 seconds and the time was spent on the storage service downloading meme content. We see three tries of `Azure.Core.Http.Request`, each of which was fast, and the time between them corresponds to the back-off interval. The last try was successful.

If you don't have `trace-id`, or perhaps if the trace was sampled out, you might be able to filter similar operations based on the context and high latency.

For example, in Jaeger, you can filter spans based on the service, span name, attributes, and duration, which helps you to find a needle in a haystack.

In some cases, you will end up with mysterious gaps – the service was up and running but spent significant time doing nothing, as shown in *Figure 1.14*:

Figure 1.14 – A request with high latency and gaps in spans

If you don't get enough data from traces, check whether there are any logs available in the scope of this span.

You might also check resource utilization metrics – was there a CPU spike, or maybe a garbage collection pause at this moment? You might find some correlation using timestamps and context, but it's impossible to tell whether this was a root cause or a coincidence.

If you have a continuous profiler that correlates profiles to traces (yes, they can do it with `Activity.Current`), you can check whether there are profiles available for this or similar operations.

We'll see how to investigate this further with .NET diagnostics tools in *Chapter 4, Low-Level Performance Analysis with Diagnostic Tools*, but if you're curious about what happened in *Figure 1.14*, the service read a network stream that was not instrumented.

Even though we talk about individual performance issues, in many cases we don't know how widespread they are, especially when we're at the beginning of an incident. Metrics and rich queries across traces can be used to find out how common a problem is. If you're on call, checking whether an issue is widespread or becoming more frequent is usually more urgent than finding the root cause.

> **Note**
> Long-tail latency requests are inevitable in distributed systems, but there are always opportunities for optimization, with caching, collocation, adjusting timeouts and the retry policy, and so on. Monitoring P95 latency and analyzing traces for long-tail issues helps you find such areas for improvement.

Performance issues

Performance problems manifest as latency or throughput degradation beyond usual variations. Assuming you fail fast or rate-limit incoming calls, you might also see an increase in the error rate for 408, 429, or 503 HTTP status codes.

Such issues can start as a slight decrease in dependency availability, causing a service to retry. With outgoing requests taking more resources than usual, other operations slow down, and the time to process client requests grows, along with number of active requests and connections.

It could be challenging to understand what happened first; you might see high CPU usage and a relatively high GC rate – all symptoms you would usually see on an overloaded system, but nothing that stands out. Assuming you measure the dependency throughput and error rate, you could see the anomaly there, but it might be difficult to tell whether it's a cause or effect.

Individual distributed traces are rarely useful in such cases – each operation takes longer, and there are more transient errors, but traces may look normal otherwise.

Here's a list of trivial things to check first, and they serve as a foundation for more advanced analysis:

- Is there an active deployment or a recent feature rollout? You can find out whether a problem is specific to instances running a new version of code using a `service.version` resource attribute. If you include feature flags on your traces or events, you can query them to check whether degradation is limited to (or started from) the requests with a new feature enabled.

- Are issues specific to a certain API, code path, or combination of attributes? Some backends, such as Honeycomb, automate this analysis, finding attributes corresponding to a higher latency or error rate.

- Are all instances affected? How many instances are alive? Attribute-based analysis is helpful here too.

- Are your dependencies healthy? If you can, check their server-side telemetry and see whether they experience problems with other services, not just yours.

  Attribute analysis can help here as well – assuming just one of your cloud storage accounts or database partitions is misbehaving, you will see it.

- Did the load increase sharply prior to the incident? Or, if your service is auto-scaled, is the auto-scaler functioning properly, and are you able to catch up with the load?

There are more questions to ask about infrastructure, the cloud provider, and other aspects. The point of this exercise is to narrow down and understand the problem as much as possible. If the problem is not in your code, investigation helps to find a better way to handle problems like these in the future and gives you an opportunity to fill the gaps in your telemetry, so next time something similar happens, you can identify it faster.

If you suspect a problem in your code, .NET provides a set of signals and tools to help investigate high CPU, memory leaks, deadlocks, thread pool starvation, and profile code, as we'll see in *Chapter 4, Low-Level Performance Analysis with Diagnostic Tools*.

Summary

Distributed systems need a new approach to observability that simplifies investigating incidents and minimizes the time to resolve issues. This approach should focus on human experience such as data visualization, the correlation across telemetry signals, and analysis automation. It requires structured, correlated telemetry signals that work together and new tools that leverage them to build a rich experience.

Distributed tracing is one such signal – it follows requests through any system and describes service operations with spans, the events representing operations in the system. .NET supports distributed tracing and integrates natively with OpenTelemetry, which is a cross-language platform to collect, process, and export traces, metrics, and logs in a vendor-agnostic way. Most modern vendors are compatible with OpenTelemetry and leverage distributed tracing capabilities. The OpenTelemetry ecosystem includes a diverse set of shared instrumentation libraries that automate common telemetry collection needs.

Distributed tracing enables correlation and causation by propagating context within the process and between services. OpenTelemetry defines standard semantics for common technologies so that vendors can build trace visualizations, application maps, shared dashboards, alerts, or queries that rely on consistent and standard attributes. Trace context and consistent attributes enable correlation between spans, logs, metrics, and any other signals coming from your system.

Individual issues can be efficiently analyzed with distributed tracing and investigations into widespread performance issues rely on attributes and timestamp correlation on metrics and across traces. Observability vendors may automate this analysis.

A combination of metrics, traces, and events gives the right number of details. Metrics allow us to receive unbiased data in a cost-effective way. By querying traces and events over high-cardinality attributes, we can answer ad hoc questions about the system.

In the next chapter, we'll get hands-on experience with distributed tracing. We'll build a demo application and explore native tracing capabilities in .NET.

Questions

1. How would you define spans and traces? What information does a span contain?
2. How does span correlation work?
3. Assuming you are on call and receive a report from a user about slow response time from your service, how would you approach the investigation?

Further reading

- *Cloud-Native Observability with OpenTelemetry* by Alex Boten

2

Native Monitoring in .NET

In this chapter, we'll explore the out-of-the-box diagnostic capabilities of modern .NET applications, starting with logs and ad hoc diagnostics, and then move on to examine what OpenTelemetry provides on top of that. We'll create a sample application and instrument it, showcasing cross-process log correlation, and learn how we can capture verbose logs with `dotnet-monitor`. Then, we'll investigate .NET runtime counters and export them to Prometheus. Finally, we'll configure OpenTelemetry to collect traces and metrics from .NET, ASP.NET Core, and Entity Framework, and check out how basic auto-instrumentations address observability needs.

The following topics are what we'll cover:

- Native log correlation in ASP.NET Core applications

- Minimalistic monitoring with .NET runtime counters

- Install OpenTelemetry and enable common instrumentations

- Tracing and performance analysis with HTTP and database instrumentations

By the end of this chapter, you'll be ready to use distributed tracing instrumentation in .NET libraries and frameworks, enable log correlation and metrics, and leverage multiple signals together to debug and monitor your applications.

Technical requirements

We're going to start building a sample application and will use the following tools for it:

- .NET SDK 7.0 or newer

- Visual Studio or Visual Studio Code with the C# development setup are recommended, but any text editor would work

- Docker and `docker-compose`

The application code can be found in the book's repository on GitHub at `https://github.com/PacktPublishing/Modern-Distributed-Tracing-in-.NET/tree/main/chapter2`.

Building a sample application

As shown in *Figure 2.1*, our application consists of two REST services and a MySQL database:

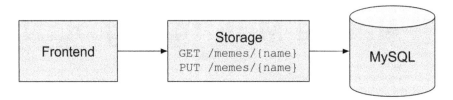

Figure 2.1 – Meme service diagram

- **Frontend**: ASP.NET Core Razor Pages application that serves user requests to upload and download images
- **Storage**: ASP.NET Core WebAPI application that uses Entity Framework Core to store images in a MySQL database or in memory for local development

We'll see how to run the full application using Docker later in this chapter. For now, run it locally and explore the basic logging and monitoring features that come with modern .NET.

We're going to use the `Microsoft.Extensions.Logging.ILogger` API throughout this book. `ILogger` provides convenient APIs to write structured logs, along with verbosity control and the ability to send logs anywhere.

ASP.NET Core and Entity Framework use `ILogger`; all we need to do is configure the logging level for specific categories or events to log incoming requests or database calls, and supply additional context with logging scopes. We're going to cover this in detail in *Chapter 8*, *Writing Structured and Correlated Logs*. For now, let's see log correlation in action.

Log correlation

ASP.NET Core enables log correlation across multiple services by default. It creates an activity that loggers can access using `Activity.Current` and configures `Microsoft.Extensions.Logging` to populate the trace context on logging scopes. ASP.NET Core and `HttpClient` also support W3C Trace Context by default, so the context is automatically propagated over HTTP.

Some logging providers, for example, OpenTelemetry, don't need any configuration to correlate logs, but our meme application uses a console provider, which does not print any logging scopes by default.

So let's configure our console provider to print scopes and we'll see the trace context on every log record. Let's also set the default level to Information for all categories just so we can see the output:

appsettings.json

```
"Logging": {
  "LogLevel" : {"Default": "Information"},
  "Console" : {"IncludeScopes" : true}
}
```

https://github.com/PacktPublishing/Modern-Distributed-Tracing-in-.NET/blob/main/chapter2/storage/appsettings.json

> **Note**
>
> Usually, you would use Information for application code only and set Warning or Error for frameworks and third-party libraries.

Let's check it out – start the storage first, then, in a different terminal, start the frontend:

```
storage$ dotnet run
frontend$ dotnet run
```

Make sure to keep both terminals open so we can check logs later. Now, let's get the preloaded meme from the frontend in your browser – hit http://localhost:5051/Meme?name=dotnet and then check the logs.

On the frontend, you may see something like this (other logs and scopes are omitted for brevity):

```
info: System.Net.Http.HttpClient.storage.LogicalHandler[101]
      => SpanId:4a8f253515db7fec, TraceId:e61779
      516785b4549905032283d93e09, ParentId:00000000000000
      00 => HTTP GET http://localhost:5050/memes/dotnet
      End processing HTTP request after 182.2564ms - 200
info: Microsoft.AspNetCore.Hosting.Diagnostics[2]
      => SpanId:4a8f253515db7fec, TraceId:e6177951678
      5b4549905032283d93e09, ParentId:0000000000000000 =>
      Request finished HTTP/1.1 GET http://localhost:
        5051/Meme?name=dotnet - 200 256.6804ms
```

The first record describes the outgoing call to the storage service. You can see the status, duration, HTTP method, and URL, as well as the trace context. The second record describes an incoming HTTP call and has similar information.

> **Note**
>
> This trace context is the same on both log records and belongs to the incoming HTTP request.

Let's see what happened on the storage:

```
info: Microsoft.AspNetCore.Hosting.Diagnostics[2]
      => SpanId:5a496fb9adf85727, TraceId:e61779516785b
      4549905032283d93e09, ParentId:67b3966e163641c4
      Request finished HTTP/1.1 GET http://localhost:
        5050/memes/dotnet - 200 1516 102.8234ms
```

> **Note**
>
> The `TraceId` value is the same on the frontend and storage, so we have cross-process log correlation out of the box.

If we had OpenTelemetry configured, we'd see a trace similar to that shown in *Figure 2.2*:

Figure 2.2 – Communication between the frontend and storage

We already know that ASP.NET Core creates an activity for each request – it reads trace context headers by default, but we can configure a different propagator. So, when we make an outgoing request from the frontend to storage, `HttpClient` creates another activity – a child of the ASP.NET Core one. `HttpClient` injects the trace context from its activity to the outgoing request headers so that they flow to the storage service, where ASP.NET Core parses them and creates a new activity, which becomes a child of the outgoing request.

Even though we didn't export activities, they are created and are used to enrich the logs with trace context, enabling correlation across different services.

Without exporting activities, we achieve correlation but not causation. As you can see in the logs, `ParentId` on storage is not the same as `SpanId` on the outgoing HTTP request.

> **Hint on causation**
>
> What happens here is that the outgoing request activity is created inside `HttpClient`, which does not write logs with `ILogger`. The log record on the outgoing request we just saw is written by the handler in the `Microsoft.Extensions.Http` package. This handler is configured by ASP.NET Core. When the handler logs that the request is starting, the `HttpClient` activity has not yet been created, and when the handler logs that the request is ended, the `HttpClient` activity is already stopped.

So, with ASP.NET Core and ILogger, we can easily enable log correlation. However, logs don't substitute distributed traces – they just provide additional details. Logs also don't need to duplicate traces.

Avoiding duplication is important: once, the author saved a company $80k a month by dropping logs that were duplicated by rich events.

Going forward in this book, we'll use logs for debugging and capturing additional information that covers gaps in traces.

Using logs in production

To record logs in production, where we have multiple instances of services and restricted access to them, we usually need a log management system – a set of tools that collect and send logs to a central location, potentially enriching, filtering, or parsing them along the way. OpenTelemetry can help us collect logs, but we also need a backend to store, index, and query the logs using any context, including `TraceId`. With this, we can easily navigate from traces to logs when needed.

On-demand logging with dotnet-monitor

It could be useful to dynamically increase log verbosity at runtime to get more detailed information while reproducing the issue or, when the log exporting pipeline is broken, get logs from the service directly.

It's possible with `dotnet-monitor` – a diagnostics tool that's able to connect to a specific .NET process and capture logs, counters, profiles, and core dumps. We'll talk about it in *Chapter 3*, *The .NET Observability Ecosystem*.

Let's install and start `dotnet-monitor` to see what it can do:

```
$ dotnet tool install -g dotnet-monitor
$ dotnet monitor collect
```

> **Note**
>
> If you're on macOS or Linux, you need to authenticate requests to the `dotnet-monitor` REST API. Please refer to the documentation at `https://github.com/dotnet/dotnet-monitor/blob/main/documentation/authentication.md` or, for demo purposes only, disable authentication with the `dotnet monitor collect -no-auth` command.

If you still have frontend and storage services running, you should see them among the other .NET processes on your machine when you open `https://localhost:52323/processes` in your browser:

```
{"pid": 27020, …, "name": "storage"},
{"pid": 29144, … "name": "frontend"}
```

Now let's capture some debug logs from storage via `dotnet-monitor` by requesting the following:

```
https://localhost:52323/logs?pid=27020&level=debug&duration=60
```

It connects to the process first, enables the requested log level, and then starts streaming logs to the browser for 60 seconds. It doesn't change the logging level in the main logging pipeline, but will return the requested logs directly to you, as shown in *Figure 2.3*:

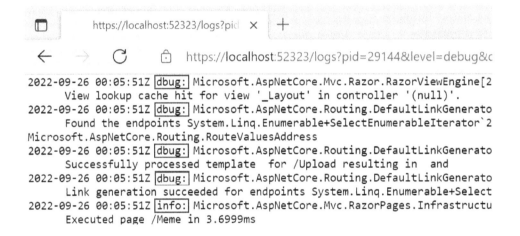

Figure 2.3 – Ad hoc logging with dynamic level using dotnet-monitor

You can apply a more advanced configuration using the POST logs API – check out `https://github.com/dotnet/dotnet-monitor` to learn more about it and other `dotnet-monitor` features.

Using `dotnet-monitor` in production on a multi-instance service with restricted SSH access can be challenging. Let's see how we can do it by running `dotnet-monitor` as a sidecar in Docker. It's also possible to run it as a sidecar in Kubernetes.

Monitoring with runtime counters

So, we have correlated logs from the platform and services with which we can debug issues. But what about system health and performance? .NET and ASP.NET Core expose event counters that can give some insights into the overall system state.

We can collect counters with OpenTelemetry without running and managing `dotnet-monitor`. But if your metrics pipeline is broken (or if you don't have one yet), you can attach `dotnet-monitor` to your process for ad hoc analysis.

`dotnet-monitor` listens to `EventCounters` reported by the .NET runtime and returns them on an HTTP endpoint in **Prometheus exposition format**.

> **Note**
>
> **Prometheus** is a metrics platform that scrapes and stores metrics. It supports multidimensional data and allows us to slice, dice, filter, and calculate derived metrics using **PromQL**.

We're going to run our service as a set of Docker containers with `dotnet-monitor` running as a sidecar for the frontend and storage, and configure Prometheus to scrape metrics from `dotnet-monitor` sidecars, as shown in *Figure 2.4*:

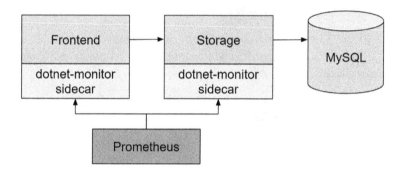

Figure 2.4 – Meme services with runtime counters in Prometheus

This makes our setup closer to real life, where we don't have the luxury of running `dotnet-monitor` on the service instance.

So, let's go ahead and run our application. Open the terminal, navigate to the `chapter2` folder, and run the following:

```
$ docker-compose -f ./docker-compose-dotnet-monitor.yml
  up --build
```

You might see some errors while MySQL is starting up. Let's ignore them for now. After a few seconds, you should be able to reach the frontend via the same URL as before.

Let's explore the CPU and memory counters published by the .NET Runtime:

- `cpu-usage` event counter (reported as `systemruntime_cpu_usage_ratio` metric to Prometheus): Represents the CPU usage as a percentage.

- `gc-heap-size` (or `systemruntime_gc_heap_size_bytes`): Represents the approximate allocated managed memory size in megabytes.

- `time-in-gc` (or `systemruntime_time_in_gc_ratio`): Represents time spent on garbage collection since the last garbage collection.

- `gen-0-gc-count`, `gen-1-gc-count`, and `gen-2-gc-count` (or `systemruntime_gen_<gen>_gc_count`): Represents the count of garbage collections in the corresponding generation per interval. The default update interval is 5 seconds, but you can adjust it. Generation sizes are also exposed as counters.

- `alloc-rate` (or `systemruntime_alloc_rate_bytes`): Represents the allocation rate in bytes per interval.

You can also find counters coming from Kestrel, Sockets, TLS, and DNS that can be useful to investigate specific issues such as DNS outages, long request queues, or socket exhaustion on HTTP servers. Check out the .NET documentation for the full list (`https://learn.microsoft.com/dotnet/core/diagnostics/available-counters`).

ASP.NET Core and `HttpClient` request counters don't have dimensions, but would be useful if you didn't have OpenTelemetry tracing or metrics and wanted to get a very rough idea about throughput and failure rate across all APIs.

Prometheus scrapes metrics from the `dotnet-monitor` metrics endpoint. We can access it ourselves to see the raw data, as shown in *Figure 2.5*:

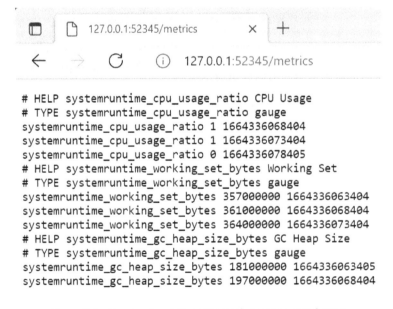

Figure 2.5 – Frontend metrics in Prometheus exposure format

It's also possible to query and plot basic visualizations with Prometheus, as you can see in *Figure 2.6*. Just hit `http://localhost:9090/graph`. For any advanced visualizations or dashboards, we would need tooling that integrates with Prometheus, such as **Grafana**.

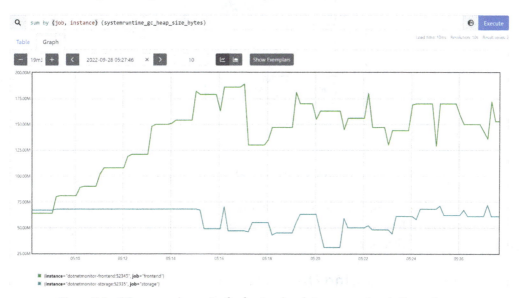

Figure 2.6 – GC memory heap size for frontend and storage services in Prometheus

As you can see, even basic ASP.NET Core applications come with minimal monitoring capabilities – counters for overall system health and correlated logs for debugging. With `dotnet-monitor` we can even retrieve telemetry at runtime without changing the code or restarting the application (well, of course, only if we have access to the application instance).

With some additional infrastructure changes to run `dotnet-monitor` as a sidecar and logging management solution, we would be able to build a very basic production monitoring solution.

We still lack distributed tracing and metrics that have rich context. Let's now see how to enable them with OpenTelemetry instrumentation and improve this experience further.

Enabling auto-collection with OpenTelemetry

In this section, we're going to add OpenTelemetry to our demo application and enable auto-collection for ASP.NET Core, `HttpClient`, Entity Framework, and runtime metrics. We'll see what it adds to the bare-bones monitoring experience.

We'll export traces to Jaeger and metrics to Prometheus, as shown in *Figure 2.7*:

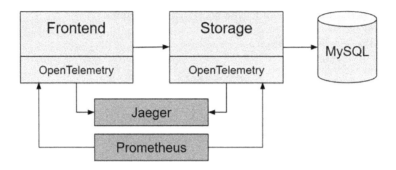

Figure 2.7 – Meme services sending telemetry to Jaeger and Prometheus

You can also send data directly to your observability backend if it has an OTLP endpoint, or you can configure a backend-specific exporter in the application. So, let's get started and instrument our application with OpenTelemetry.

Installing and configuring OpenTelemetry

OpenTelemetry comes as a set of NuGet packages. Here are a few that we're using in our demo app:

- `OpenTelemetry`: The SDK that contains all that we need to produce traces and metrics and configure a generic processing and export pipeline. It does not collect any telemetry on its own.

- `OpenTelemetry.Exporter.Jaeger`: This package contains a trace exporter that publishes spans to Jaeger.

- `OpenTelemetry.Exporter.Prometheus.AspNetCore`: This package contains the Prometheus exporter. It creates a new `/metrics` endpoint for Prometheus to scrape metrics from.

- `OpenTelemetry.Extensions.Hosting`: This package simplifies OpenTelemetry configuration in ASP.NET Core applications.

- `OpenTelemetry.Instrumentation.AspNetCore`: This package enables ASP.NET Core tracing and metrics auto-instrumentation.

- `OpenTelemetry.Instrumentation.Http`: This package enables tracing and metrics auto-instrumentation for `System.Net.HttpClient`.

- `OpenTelemetry.Instrumentation.EntityFrameworkCore`: Tracing instrumentation for Entity Framework Core. We only need it for the storage service.

- `OpenTelemetry.Instrumentation.Process` and `OpenTelemetry.Instrumentation.Runtime`: These two packages enable process-level metrics for CPU and memory utilization and include the runtime counters we saw previously with `dotnet-monitor`.

You can also enable other counter sources one by one with the `OpenTelemetry.Instrumentation.EventCounters` package.

Distributed tracing

To configure tracing, first call `AddOpenTelemetry` extension method on `IServiceCollection` and then call the `WithTracing` method, as shown in the following example:

Program.cs

```
builder.Services.AddOpenTelemetry().WithTracing(
  tracerProviderBuilder => tracerProviderBuilder
    .AddJaegerExporter()
    .AddHttpClientInstrumentation()
    .AddAspNetCoreInstrumentation()
    .AddEntityFrameworkCoreInstrumentation());
```

https://github.com/PacktPublishing/Modern-Distributed-Trac-ing-in-.NET/blob/main/chapter2/storage/Program.cs

---

Here, we're adding the Jaeger exporter and enabling `HttpClient`, ASP.NET Core, and Entity Framework instrumentations (on storage).

We're also configuring the service name via the `OTEL_SERVICE_NAME` environment variable in `launchSetting.json` and in `docker-compose-otel.yml` for Docker runs. The OpenTelemetry SDK reads it and sets the `service.name` resource attribute accordingly.

The Jaeger host is configured with the `OTEL_EXPORTER_JAEGER_AGENT_HOST` environment variable in `docker-compose-otel.yml`.

We'll talk more about configuration in *Chapter 5*, *Configuration and Control Plane*, and learn how to configure sampling, enrich telemetry, and add custom sources.

Metrics

Metrics configuration is similar – we first call the `AddOpenTelemetry` extension method on `IServiceCollection` and then in the `WithMetrics` callback set up the Prometheus exporter and auto-instrumentations for `HttpClient`, ASP.NET Core, Process, and Runtime. Entity Framework's instrumentation does not report metrics.

Program.cs

```
builder.Services.AddOpenTelemetry()
        ...
        .WithMetrics(meterProviderBuilder => meterProviderBuilder
            .AddPrometheusExporter()
            .AddHttpClientInstrumentation()
            .AddAspNetCoreInstrumentation()
            .AddProcessInstrumentation()
            .AddRuntimeInstrumentation());
```

https://github.com/PacktPublishing/Modern-Distributed-Trac-
ing-in-.NET/blob/main/chapter2/storage/Program.cs

We also need to expose the Prometheus endpoint after building the application instance:

Program.cs

```
var app = builder.Build();
app.UseOpenTelemetryPrometheusScrapingEndpoint();
```

https://github.com/PacktPublishing/Modern-Distributed-Trac-
ing-in-.NET/blob/main/chapter2/storage/Program.cs

We're ready to run the application!

```
$ docker-compose -f docker-compose-otel.yml up --build
```

You should see logs from all services including some errors while MySQL is starting up. Check the frontend to make sure it works: `https://localhost:5051/Meme?name=dotnet`.

Exploring auto-generated telemetry

The meme service is now up and running. Feel free to upload your favorite memes and if you see any issues, use telemetry to debug them!

Debugging

If you try to upload something right after the service starts, you might get an error like the one shown in *Figure 2.8*. Let's figure out why!

Figure 2.8 – Error from application with traceparent

We can approach this investigation from two angles. The first is to use the `traceparent` shown on the page; the second is to filter the traces from the frontend based on the error status. In any case, let's go to Jaeger – our tracing backend running on `http://localhost:16686/`. We can search by `Trace ID` or filter by service and error, as shown in *Figure 2.9*:

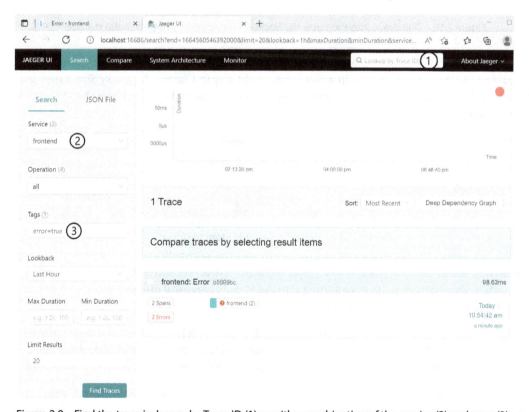

Figure 2.9 – Find the trace in Jaeger by Trace ID (1) or with a combination of the service (2) and error (3)

If we open the trace, we'll see that the storage service refused the connection – check out *Figure 2.10*. What happened here?

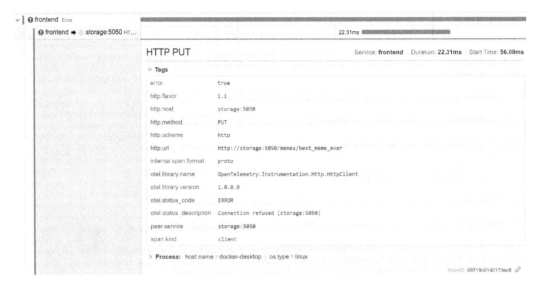

Figure 2.10 – Drill down into the trace: the frontend service could not reach the storage

Since there are no traces from the storage, let's check the storage logs with `docker logs chapter2-storage-1`. We'll query the logs in a more convenient way in *Chapter 8, Writing Structured and Correlated Logs*. For now, let's just grep storage logs around the time the issue occurred and find the relevant record, as shown in *Figure 2.11*:

```
Body: An error occurred using the connection to database '{database}' on server '{server}'.
Attributes:
    -> dotnet.ilogger.category: STRING(Microsoft.EntityFrameworkCore.Database.Connection)
    -> database: STRING()
    -> server: STRING(mysql)
    -> Id: INT(20004)
    -> Name: STRING(Microsoft.EntityFrameworkCore.Database.Connection.ConnectionError)
```

Figure 2.11 – Connection error in storage stdout

Apparently, the storage was not able to connect to the MySQL server and it could not start until the connection was established. If we dig further into the MySQL logs, we'll discover that it took a while for it to start, but then everything worked just fine.

Some action items from this investigation are to enable retries on the frontend and investigate the slow start for MySQL. If it happens in production where there are multiple instances of storage, we should also dig into the load balancer and service discovery behavior.

What tracing brings here is *convenience* – we could have done the same investigation with logs alone, but it would have taken longer and would be more difficult. Assuming we dealt with a more complicated case with dozens of requests over multiple services, parsing logs would simply not be a reasonable option.

As we can see in this example, tracing can help us narrow the problem down, but sometimes we still need logs to understand what's going on, especially for issues during startup.

Performance

Let's check out some metrics collected by the HTTP and runtime instrumentations.

OpenTelemetry defines `http.server.duration` and `http.client.duration` **histogram** metrics with low-cardinality attributes for method, API route (server only), and status code. These metrics allow us to calculate latency percentiles, throughputs, and error rates.

With OpenTelemetry metrics, ASP.NET Core instrumentation can populate API routes so we can finally analyze latency, throughput, and error rate per route. And histograms give us even more flexibility – we can now check the distribution of latency rather than the median, average, or a predefined set of percentiles.

Latency

HTTP client latency can be defined as the time between initiating a request and the response being received. For servers, it's the time between receiving a request and the end of the server's response.

> **Tip**
>
> When analyzing latency, filter out errors and check the distribution of latency rather than just averages or medians. It's common to check the 95th percentile (aka P95).

Figure 2.12 shows P95 latency for the `PUT /meme` API on the client and server side:

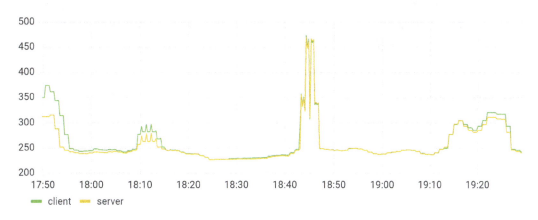

Figure 2.12 – Server versus client PUT /meme latency P95 in milliseconds

Time to first byte versus time to last byte

In .NET, `HttpClient` buffers a response before returning it, but it can be configured to return the response right after the headers are received with `HttpCompletionOptions`. `HttpClient` instrumentation can't measure time-to-last-byte in this case.

The distinction between *time to first body byte* versus *time to last byte* can be important on frontends with clients using unreliable connections or when transferring a lot of data. In such cases, it's useful to instrument stream operations and then measure the time to first byte *and* time to last byte. You can use the difference between these metrics to get an idea about connection quality and optimize the end user experience.

Error rate

Error rate is just a rate of unsuccessful requests per a given period of time. The key question here is what constitutes an error:

- `1xx`, `2xx`, and `3xx` status codes indicate success.
- `5xx` codes cover errors such as the lack of a response, a disconnected client, network, or DNS issues.
- Status codes in the `4xx` range are hard to categorize. For example, `404` could represent an issue – maybe the client expected to retrieve the data but it's not there – but could also be a positive scenario, where the client is checking whether a resource exists before creating or updating it. There are similar concerns with other statuses.

> **Note**
>
> OpenTelemetry only marks client spans with `4xx` as errors. We'll see in *Chapter 5, Configuration and Control Plane*, how to tailor it to your needs.

It's also common to treat latency above a given threshold as an error to measure availability, but we don't strictly need it for observability purposes.

Figure 2.13 shows an example of a server error rate chart for a single API grouped by error code:

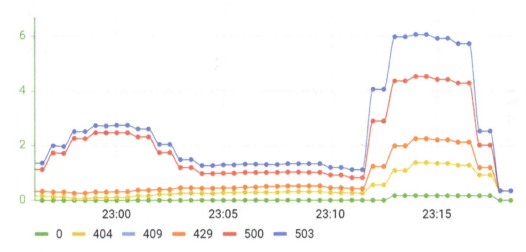

Figure 2.13 – Error rate per second for the GET/meme API grouped by error code

It is also important to calculate the error rate per API route and method on servers. Because of different request rates, it's easy to miss spikes or changes in less frequently called APIs.

> **Tip**
> Returning precise status codes for "known" errors and letting the service return 500 only for unhandled exceptions makes it easier to use your service, but also simplifies monitoring and alerting. By looking at the error code, we can discern the possible reasons and not waste time on known cases. Any 500 response becomes important to investigate and fix or handle properly.

To check resource consumption, we can use runtime and process metrics. For example, *Figure 2.14* shows CPU usage:

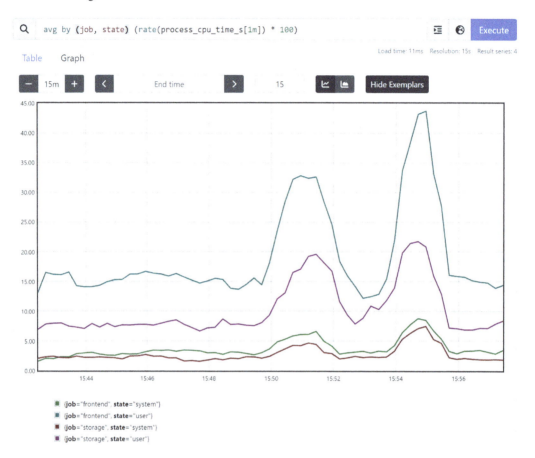

Figure 2.14 – CPU usage query

The query returns the average CPU utilization percentage across all instances for each service represented by job dimension – we configured jobs in `configs/prometheus-otel.yml`.

The state dimension divides processor time into user and privileged (system) time. To calculate the total average CPU usage per instance per service, we could write another Prometheus query:

```
sum by (job, instance) (rate(process_cpu_time_s[1m]) * 100)
```

The query calculates the total CPU usage per instance and then calculates the average value per service.

As you can see, the Prometheus query language is a powerful tool allowing us to calculate derived metrics and slice, dice, and filter them.

We'll see more examples of runtime metrics and performance analysis in *Chapter 4, Low-Level Performance Analysis with Diagnostic Tools.*

Summary

In this chapter, we explored .NET diagnostics and monitoring capabilities supported by the platform and frameworks. ASP.NET Core context propagation is enabled by default and logging providers can use it to correlate logs. We need a log management system to be able to store logs from multiple instances of a service and efficiently query them.

`dotnet-monitor` allows the streaming of logs on demand from specific instances of your service, and the scraping of event counters with Prometheus to get a basic idea about service health. It can also be used for low-level performance analysis and can be run in production.

Then, we enabled OpenTelemetry auto-instrumentation for the HTTP stack and Entity Framework. HTTP and DB traces enable basic debugging capabilities, providing generic information on what happened for each remote call. You can search for traces based on attributes and query them using your tracing backend. With tracing, we can easily find a problematic service or component, and when that's not enough, we can retrieve logs to get more details about the issue. With logs correlated to traces, we can easily navigate between them.

HTTP metrics enable common performance analysis. Depending on your backend, you can query, filter, and derive metrics and build dashboards and alerts based on them.

Now that we've got hands-on experience with basic distributed tracing and metrics, let's explore the .NET ecosystem more and see how you can leverage instrumentation for common libraries and infrastructure.

Questions

1. How would you show trace context on a Razor page?

2. Imagine that the observability backend stopped receiving telemetry from some instances of the service. What can we do to understand what's going on with these instances?

3. With the help of the Prometheus documentation (`https://prometheus.io/docs/prometheus/latest/querying/basics/`), write a query with PromQL to calculate the throughput (requests per second) per service and API.

4. With our meme service, how would you find out when a meme was uploaded and how many times it had been downloaded if you know only the meme's name?

3

The .NET Observability Ecosystem

In the previous chapter, we explored .NET observability features included into the platform and frameworks, but there are more instrumentations covering other libraries and environments.

In this chapter, we'll learn how to find and evaluate instrumentations and then take a closer look at instrumentations for a few specific libraries: StackExchange.Redis, Azure, and AWS SDKs. We'll also explore tracing and metrics coming from infrastructure using **Dapr** (**distributed application runtime**) as an example. Finally, we'll see how to configure tracing in serverless environments where we have less control, but observability is even more important.

Through this chapter, you'll learn:

- How to find, evaluate, and enable OpenTelemetry instrumentations
- What Dapr and service meshes are capable of when it comes to observability
- How to enable tracing in serverless environments

By the end of this chapter, you'll get hands-on experience with different kinds of instrumentations and you will be able to configure and use distributed tracing for a wide range of backend applications. Let's get started!

Technical requirements

In this chapter, we're going to evolve our meme application and use a cloud object store, Amazon S3 or Azure Blob Storage, along with a local Redis cache. The code for this chapter is available in the book's GitHub repository at `https://github.com/PacktPublishing/Modern-Distributed-Tracing-in-.NET/tree/main/chapter3`, which has the following folder structure:

- `libraries`: Contains library instrumentation sample app for the first section of this chapter
- `dapr`: Contains Dapr instrumentation sample for the second section

- serverless: Contains aws and azure folders with examples of AWS Lambda and Azure Functions instrumentations

To run these applications, you would need the following tools:

- .NET SDK 7.0 or later
- Visual Studio or VS Code, but any text editor would work
- Docker and docker-compose
- Dapr CLI
- An Azure subscription (optional):
  - We're going to use Blob Storage and Application Insights.
  - With Blob Storage, we're going to stay well within free-tier limits. Application Insights does not have a free tier, but you can still try it out with Azure promotional credits.
  - We'll use Azure Function Tools v4 and (optionally) Azure CLI.
- An AWS subscription (optional):
  - We're going to use S3, Lambda, and X-Ray. We'll stay well within free-tier limits for each of them.
  - We'll need AWS toolkit for VS or Lambda .NET CLI and (optionally) AWS CLI.

Configuring cloud storage

If you don't want to create an Azure or AWS subscription, you can still run libraries and dapr samples locally by setting CloudStorage.Type to Local in storage/appsettings.json. There is no local setup for serverless demos.

Otherwise, set CloudStorage.Type to the storage of your choice, AwsS3 or AzureBlob, and let's see how to configure them.

AWS S3

Create a new bucket using AWS console or CLI:

```
$ aws s3api create-bucket -bucket <name> --region <region>
```

Then, add bucket info to libraries/storage/appsettings.json.

We'll also need credentials to access blob storage and we're going to use the credentials file where we can. You can generate one using the aws configure command. Applications would search for AWS credentials file at ${HOME}/.aws/credentials.

Replace the HOME environment variable in `docker-compose.yml` in the `libraries/` and `serverless/aws` folders.

Azure Blob Storage

Create a new storage account. You can use Azure portal or Azure CLI and then obtain a connection string:

```
$ az storage account create --resource-group <group> --name
<account>
$ az storage account show-connection-string --resource-
group <group> --name <account>
```

Add the connection string to `.env` file next to `libraries/docker-compose.yml` in the following format:

```
AZURE_BLOB_CONNECTION_STRING="DefaultEndpointsProtocol=
    https;...."
```

Using instrumentations for popular libraries

In the previous chapter, we saw how to enable tracing for the .NET platform, ASP.NET Core, and Entity Framework to cover the basics, but anyone can create instrumentation for a popular library and share it with the community. Also, with tracing and metrics primitives being part of .NET and OpenTelemetry to collect data in a vendor-agnostic way, libraries can add native instrumentation.

There are multiple terms that describe different kinds of instrumentations:

- **Auto-instrumentation** *sometimes* implies that instrumentation can be enabled without *any* modification of application code, but is sometimes used to describe any shared instrumentation that is easy to enable.

- **Instrumentation library** means that you can enable instrumentation by installing the corresponding NuGet package and configuring it with a few lines of code at startup time.

- **Native instrumentation** implies that instrumentation code is a part of the library, so no additional NuGet package is necessary, but you may still need to enable instrumentation.

- **Manual instrumentation** is the one that you write yourself as a part of your application code.

The boundaries between automatic, native, and instrumentation libraries are blurry. For example, the HTTP client contains native instrumentation starting with .NET 7.0, but you might still enable it in a more convenient way with the corresponding instrumentations. Or, with some bytecode rewrite that configures OpenTelemetry, we can enable library instrumentations without changing any of the application code. In this book, we use a relaxed version of the auto-instrumentation term (for the lack of a better one) to describe all non-manual instrumentations, but we mention a specific kind when it's relevant.

There are several sources where we can find available instrumentations:

- **OpenTelemetry registry** (`https://opentelemetry.io/registry`): You can filter the instrumentations by language and component. Many instrumentations are not added to the registry though. It lists all kinds of instrumentation regardless of their kind.

- **OpenTelemetry .NET repo** (`https://github.com/open-telemetry/opentelemetry-dotnet`): Contains library instrumentation for .NET frameworks and libraries. The ASP.NET Core and HTTP client instrumentations we used in the previous chapter live here along with SQL, gRPC, and exporters for OSS backends. These are instrumentation libraries.

- **OpenTelemetry Contrib repo** (`https://github.com/open-telemetry/opentelemetry-dotnet-contrib`): Contains different OpenTelemetry components: instrumentation libraries, exporters, and other utilities. You can find instrumentations for AWS SDK, ElasticSearch, WCF, StackExchange.Redis, and more there. The Entity Framework instrumentation we used in the previous chapter also lives in this repo.

- **OpenTelemetry instrumentation repo** (`https://github.com/open-telemetry/opentelemetry-dotnet-instrumentation`): Contains fully codeless auto-instrumentations that work via different mechanism - .NET profiling API. You can find GraphGL and MongoDB instrumentation there. In addition to auto-instrumentations for specific libraries; it provides a mechanism to configure OpenTelemetry in a codeless way that includes a set of common instrumentation libraries.

- **Other sources**: If you didn't find what you're looking for in the registry or the OpenTelemetry repos, search for issues in OpenTelemetry repos and don't forget to check your library repo. For example, you can find MongoDB instrumentation at `https://github.com/jbogard/MongoDB.Driver.Core.Extensions.DiagnosticSources`, which is leveraged in the *instrumentation* repo but can be used as a standalone instrumentation library.

When adding instrumentations, pay attention to their stability and maturity. Instrumentations in the `opentelemetry-dotnet` repo are widely used but are not yet stable (it could have changed by the time you read this).

Instrumentations in the *contrib* repo have different statuses; for example, AWS is stable, while MySQL is in alpha and works for relatively old versions of the `MySQL.Data` package at the time of writing.

> **Tip**
> If you decide to take dependency on a less common preview package, make sure to test it well. Compatibility with your version of the client library, stability, and performance should be the main concerns. All of them should be covered with integration and stress-testing—just make sure to enable instrumentation!

It's good to get a basic idea of how the instrumentation works and check whether the mechanism behind it satisfies your performance requirements. For example, native instrumentations rely on `ActivitySource` or `DiagnosticSource`, and MongoDB and AWS instrumentations rely on hooks in corresponding libraries. All of these methods should work reasonably well, but the `MySQL.Data` instrumentation relies on `System.Diagnostics.TraceListener`, which is not thread-safe by default, and, when configured to be thread-safe, is not performant.

Even the most efficient instrumentations come with some performance hit. You should expect throughput to drop a few percent compared to non-instrumented code. Specific numbers heavily depend on your scenarios and OpenTelemetry configuration, such as sampling.

> **Note**
>
> Many developers consider auto-instrumentations to be magical and avoid them for this reason. By learning the mechanisms behind instrumentation, you can identify areas for additional testing, understand limitations, and gain confidence to use it (or not).

So, let's instrument the new version of the meme service and dig deep into each instrumentation we're going to use.

Instrumenting application

Our new demo application stores memes in Azure Blob Storage or AWS S3 and caches them in Redis, as shown in *Figure 3.1*:

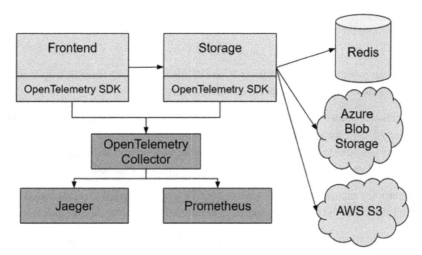

Figure 3.1 – Meme service with configurable cloud storage

You can also set it up to store memes in Redis if you don't want to configure a cloud subscription.

There are no changes on **frontend** from the previous chapter—we already enabled OpenTelemetry with HTTP instrumentations there. On **storage**, though we still need to add a few more instrumentations for AWS, Redis, and Azure SDK.

First, we need to install `OpenTelemetry.Contrib.Instrumentation.AWS` and `OpenTelemetry.Instrumentation.StackExchangeRedis` and then configure them:

libraries\storage\Program.cs

```
builder.Services.AddOpenTelemetry()
  .WithTracing(tracerProviderBuilder =>
        tracerProviderBuilder.
      .AddRedisInstrumentation(redisConnection, o =>
           o.SetVerboseDatabaseStatements = true)
      .AddAWSInstrumentation(o =>
           o.SuppressDownstreamInstrumentation = false)
      ...);
```

https://github.com/PacktPublishing/Modern-Distributed-Tracing-in-.NET/blob/main/chapter3/libraries/storage/Program.cs

Let's unpack it and explore instrumentations one by one.

Redis

Redis instrumentation is available via the `OpenTelemetry.Instrumentation.StackExchangeRedis` package and comes from the *contrib* repo—documentation and examples are available there.

Let's see how we can evaluate this instrumentation. While any details about it might change, the approach can be applied to any other instrumentation library.

Redis instrumentation is not stable at the time of writing but it has a fair number of downloads on NuGet and no bugs reported. If we investigate how it works, we'll see that it leverages the `StackExchange.Redis` profiling APIs—hooks allowing the start of a profiling session and recording events that happen during its execution. Despite the name, it doesn't need the profiler attached.

It's a relatively complex instrumentation—a profiling API is not designed for distributed tracing, so instrumentation must cover the gaps by maintaining an internal cache of sessions and cleaning them up.

To enable instrumentation, we call the `AddRedisInstrumentation` extension method on `TracerProviderBuilder` and pass the connection instance. If you have more than one connection, you'll have to enable instrumentation for each of them.

We also passed instrumentation options and enabled verbose database statements to collect additional data including Redis keys and scripts by setting `SetVerboseDatabaseStatements` flag to `true`:

```
AddRedisInstrumentation(redisConnection, o =>
  o.SetVerboseDatabaseStatements = true)
```

It's a good idea to check how this configuration might affect application performance and the verbosity of the output before deploying it to production. If we look into the Redis instrumentation code, this flag guards reflection-based (but efficient) calls to obtain the command key and script.

Depending on what we store in Redis, we should also make sure it does not record any secrets or sensitive data.

You probably noticed that instrumentations follow a common pattern, but unlike Redis ones, most of them are global and don't require a per-client instance setup.

There are other options that control tracing on Redis: you can specify callback to enrich activities, disable events with additional timings, and configure intervals to clean up completed profiling sessions.

If we start the application now and upload and download several memes on `http://localhost:5051/`, we'd see traces like the one shown in *Figure 3.2* for meme download flow:

Figure 3.2 – Meme download with Redis span

You can see the standard `net.peer.*` attributes describing generic network endpoint and `db.*` attributes describing database call with `db.statement` matching Redis command and key. We only see the key (`this_is_fine`) since we set `SetVerboseDatabaseStatements` to `true`, otherwise `db.statement` would match the command `HMGET`.

You can also see three logs (span events in Jaeger) describing additional timings for the Redis command. Since Redis is quite fast, you might find these events to be not very useful and disable them by setting `EnrichActivityWithTimingEvents` to `false`, which should decrease your observability bill and slightly improve performance.

AWS SDK

AWS SDK instrumentation is available in the `OpenTelemetry.Contrib.Instrumentation.AWS` NuGet package with the code residing in the *contrib* repo. Let's try to evaluate it using the same approach.

It is stable and relies on a global tracing handler that applies to all AWS clients and instances, not just S3. This handler in turn leverages .NET tracing primitives: `Activity` and `ActivitySource`.

To enable AWS instrumentation, just call the `AddAWSInstrumentation` extension method on `TracerProviderBuilder`. At this moment, there's just one configurable option that controls whether nested HTTP calls should be traced:

```
AddAWSInstrumentation(o => o
    .SuppressDownstreamInstrumentation = false)
```

Figure 3.3 shows the meme upload trace: **frontend** calls **storage** and **storage** calls `PutObject` that in turn makes an HTTP PUT request to S3. After the meme is uploaded, it's cached on Redis:

Figure 3.3 – Upload meme to S3

The nested HTTP span is coming from the HTTP Client instrumentation, and we only see it because `SuppressDownstreamInstrumentation` is set to `false`.

If we expand `S3.PutObject`, we'll see attributes that describe this operation, as shown in *Figure 3.4*:

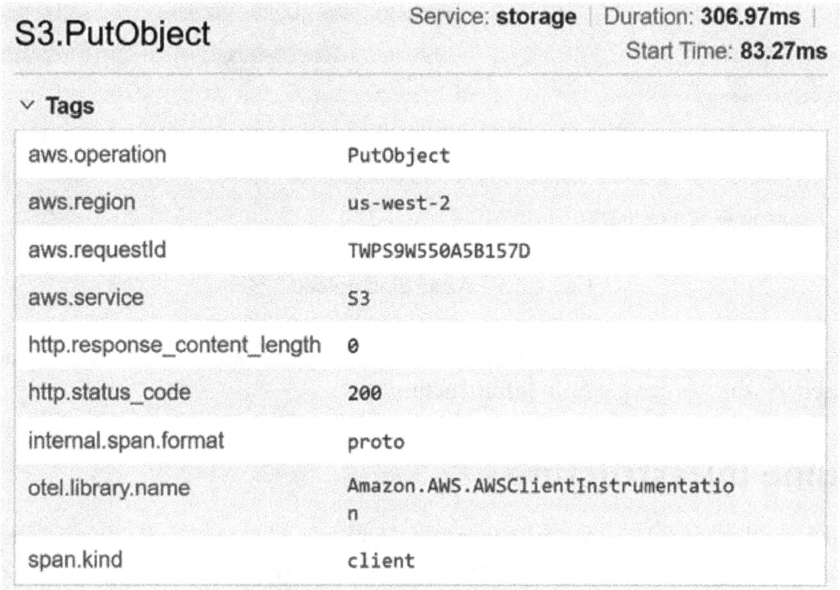

Figure 3.4 – AWS S3 span attributes

Azure SDK

Azure SDK instrumentation is native—it's baked into modern libraries—and you don't need to install any additional packages. Tracing code for all client libraries is available in the `https://github.com/Azure/azure-sdk-for-net/` repo. Still, it's not stable because of tracing semantic conventions being experimental. For example, attribute names, types, and relationships between activities may change in the future.

You can enable it with `AppContext` switch either in `csproj` or by adding the following code before Azure clients' initialization:

```
AppContext.SetSwitch(
  "Azure.Experimental.EnableActivitySource",
  true)
```

Instrumentation uses `ActivitySource` and `Activity` directly, so all we need to enable it is to call the `AddSource("Azure.*")` method on `TracerProviderBuilder`. It enables all sources that start with `Azure`, but you can also enable individual sources.

Figure 3.5 shows the Azure SDK blob upload trace—logical upload operation and nested HTTP request. We see one there, but for chunked downloads, complex calls, or in case of retries, we'd see multiple nested HTTP calls:

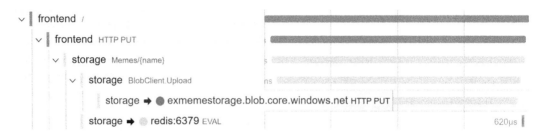

Figure 3.5 – Azure Blob upload

We explored tracing for several libraries and learned how to discover and evaluate instrumentations. Let's now discover what we can get from infrastructure.

Leveraging infrastructure

In this section, we'll explore Dapr for microservices. Dapr provides service discovery, component bindings, secret management, locking, state management, observability, and more building blocks helping developers to focus on application logic. We'll focus on distributed tracing.

In our demo application, we're going to handle all network calls with Dapr and enable tracing and metrics on it. We'll also keep telemetry enabled on the microservices. *Figure 3.6* shows the new application layout:

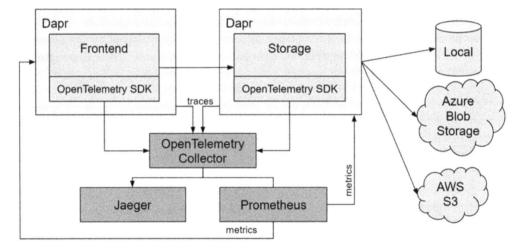

Figure 3.6 – Meme application with Dapr runtime

Dapr runs as a sidecar—a separate process wrapping each application instance. **Frontend** in our setup calls into **storage** via Dapr, which handles service discovery, error handling, encryption, load balancing, and more. **Storage**, in turn, uses Dapr output **binding** to communicate to Azure, AWS, or store memes locally.

Dapr integrates well with Kubernetes, but we'll use self-hosted mode and `docker-compose` to keep things simple.

Dapr supports distributed tracing and metrics for incoming and outgoing calls that applications make through Dapr. Let's see what it means in practice.

Configuring secrets

Dapr secrets configuration needs a different approach than we used for the libraries demo. We'll need to update `darp/configs/dapr/storage-components/secrets.json` as follows:

- For AWS, put your access keys in `{ "awsKey": <key>, "awsSecret": <secret>}`.

- For Azure, set `{ "azStorageAccount": <account>, "azStorageKey": <key>}`. If you don't have Azure credentials, remove the `binding-azure.yaml` file from the `dapr/configs/dapr/storage-components` folder, otherwise samples will not work.

- For local runs, set `CloudStorage.Type` to `Local` in `storage/appsettings.json`.

Configuring observability on Dapr

To enable tracing and metrics, let's add corresponding sections to `Configuration spec`:

./dapr/configs/dapr/config.yml

```
spec:
  metric:
    enabled: true
  tracing:
    samplingRate: "1"
    zipkin:
      endpointAddress: "http://otelcollector:9412/
        api/v2/spans"
```

https://github.com/PacktPublishing/Modern-Distributed-Tracing-in-.NET/blob/main/chapter3/dapr/configs/dapr/config.yml

We also added Dapr sidecars to `docker-compose.yml`, enabled the Zipkin trace receiver on the OpenTelemetry collector, and added Dapr metrics endpoints to Prometheus targets to scrape from. As a result, we receive traces and metrics from the application and Dapr at the same time. Let's check them out.

Tracing

Let's run the application now with `docker-compose up --build`, hit **frontend** at `http://localhost:5051`, and upload some memes. If you open Jaeger at `http://localhost:16686` and find some upload requests, you should see something like the trace shown in *Figure 3.7*:

Figure 3.7 – Trace from the application and Dapr

The first two spans coming from **frontend** didn't really change as compared to the trace we saw in *Figure 3.2* when the application didn't use Dapr—they are still coming from it directly. Then we see the `frontend /memes/d8...` and `CallLocal/storage/memes/d8...` spans—they are new and are coming from Dapr.

If we expand them as shown in *Figure 3.8*, we'll also see the attributes it set:

Figure 3.8 – Dapr spans and attributes

You would probably wonder if we still need distributed tracing on the service—let's check it.

Stop containers and comment out the OTEL_EXPORTER_OTLP_ENDPOINT environment variable in docker-compose.yml for **frontend** and **storage**; we don't enable OpenTelemetry if the endpoint is not provided.

Then, restart the application and upload some memes again, and the result is shown in *Figure 3.9*:

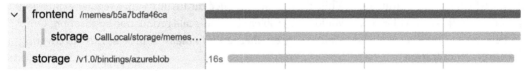

Figure 3.9 – Dapr tracing without OpenTelemetry enabled in the application

So, we see the spans coming from Dapr, but the trace does not look right—upload to Azure Blob is not a child of an incoming request represented with CallLocal/storage span. What happened there?

In *Chapter 2*, *Native Monitoring in .NET*, we have shown that ASP.NET Core and HttpClient in .NET create activities regardless of OpenTelemetry presence. This is what happened here—**frontend** and **storage** created all the spans we saw previously in *Figure 3.7*, but they are not recorded anywhere. So, CallLocal is a grandparent to /v1.0/bindings/azureblob, but the span between them is not recorded and causation is lost.

Similarly, if you use Dapr on an application that does not enable distributed tracing by default, the context will not be propagated within the **storage** service and any correlation between CallLocal and /v1.0/bindings/azureblob would disappear.

> **Note**
> Dapr or service mesh, such as Istio, can trace network calls, but they cannot propagate trace context within the application process and rely on applications to do it. They also can't stamp context on the logs if your application does not do it.

If you can't instrument your application, traces coming from Dapr or service mesh are still handy, despite being semi-correlated.

If you use Dapr for reasons beyond observability and your application is instrumented, then Dapr tracing gives you observability into Dapr itself to see how it handles requests, so you can compare latencies, debug configuration issues, and so on.

Metrics

Dapr reports extensive metrics about application communication and bindings such as HTTP and gRPC request count, duration, and request and response size histograms. You could also find Go runtime stats for the Dapr itself.

These metrics look quite promising but by default they use the HTTP request path as an attribute on metrics, which has high cardinality. While they allow to reduce cardinality with a regular expression and convert path to an API route, it would be a problem in high-scale production application. Once they become production ready, they could be a great alternative to many in-process metrics covering network communication.

Instrumenting serverless environments

Serverless environments need observability more than other systems—they are frequently used to integrate different services with little-to-no user code, making debugging and local testing difficult. With load balancing, scaling, and other common infrastructure pieces handled for us, we still need to understand what's going on when things don't work as expected.

In addition, as users, we are very limited with telemetry collection options—we can't install agents, configure runtime, or run something in privileged mode—we can only use what cloud providers expose. At the same time, cloud providers have a great opportunity to instrument code for us. Let's see what AWS Lambda and Azure Functions provide out of the box and what we can do on top of it.

AWS Lambda

AWS Lambda supports invocation tracing with X-Ray out of the box; you just need to enable active tracing via console or CLI to trace incoming calls to your function and see basic invocation metrics:

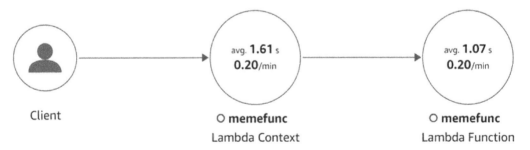

Figure 3.10 – AWS X-Ray service map showing default Lambda instrumentation

To go further than this and trace what happens in your code, you'd need to use X-Ray SDK as a stable solution or OpenTelemetry, which is in beta at this point. We're going to play with OpenTelemetry in this demo.

The configuration around OpenTelemetry is likely to change. So, we will kindly ask you to check out the latest instructions for **ADOT Collector** (**AWS Distro for OpenTelemetry Collector**), available at `https://aws-otel.github.io/docs/getting-started/lambda/lambda-dotnet`.

ADOT Collector is based on OpenTelemetry Collector; it's also compatible with AWS environments and comes with a preselected set of community components. We're going to send traces to X-Ray, which is a default configuration for ADOT Collector, but you can configure it to send data to your observability backend.

Now we're ready to explore the tracing experience in Lambda.

Enabling additional tracing

Tracing configuration in Lambda is like any other service. First, we need to install the `OpenTelemetry.Instrumentation.AWSLambda` NuGet package and then configure it along with the exporter and other instrumentations:

Function.cs

```
static Function()
{
  Sdk.SetDefaultTextMapPropagator(new
     AWSXRayPropagator());
    TracerProvider = Sdk.CreateTracerProviderBuilder()
      .AddAWSLambdaConfigurations()...;
}
```

`https://github.com/PacktPublishing/Modern-Distributed-Tracing-in-.NET/blob/main/chapter3/serverless/aws/memefunc/Function.cs`

Let's unpack what happens here. First, we set `AWSXRayPropagator` as a default context propagator—it enables context propagation over the `X-Amzn-Trace-Id` header.

Then, we enabled Lambda instrumentation with `AddAWSLambdaConfigurations`. If we look under the hood, this method does a couple of things:

- Detects and configures resource attributes such as cloud provider, region, function name, and version
- Enables `ActivitySource` that reports Lambda invocations and stitches context

Note that we do it in the static constructor to optimize performance and reduce costs. Despite being serverless, Lambda uses one process for multiple invocations.

As the last step, we need to implement the tracing handler that wraps our Lambda logic:

Function.cs

```
async Task<APIGatewayProxyResponse> TracingHandler(
  APIGatewayHttpApiV2ProxyRequest req, ILambdaContext ctx)
    =>
    await AWSLambdaWrapper.TraceAsync(TracerProvider,
      MemeHandler, req, ctx);
```

```
https://github.com/PacktPublishing/Modern-Distributed-Trac-
ing-in-.NET/blob/main/chapter3/serverless/aws/memefunc/Function.
cs
```

Note that we configured Lambda to invoke `TracingHandler` instead of inner `MemeHandler`.

If we get back to the configuration, the rest enables AWS SDK and HTTP Client instrumentation. We also configured the OTLP exporter without parameters—it uses the default endpoint (`localhost:4317`) and the default protocol (`gRPC`).

We also configured **frontend** to send data to ADOT with the X-Ray exporter, so we get all traces in the same place.

If you didn't deploy your Lambda function yet, deploy it now, for example, with AWS Toolkit for Visual Studio or Lambda tools for .NET CLI.

Make sure to configure the function URL on **frontend** as the `Storage__Endpoint` environment variable—you can set it in `./frontend/docker-compose.yml`. We don't use authorization in the demo, but make sure to secure your real-life applications.

Now, let's start **frontend** with `docker-compose up --build`, then upload and download some memes at `http://localhost:5051`.

Let's switch to AWS X-Ray and check out the traces. You should see something similar to *Figure 3.11*:

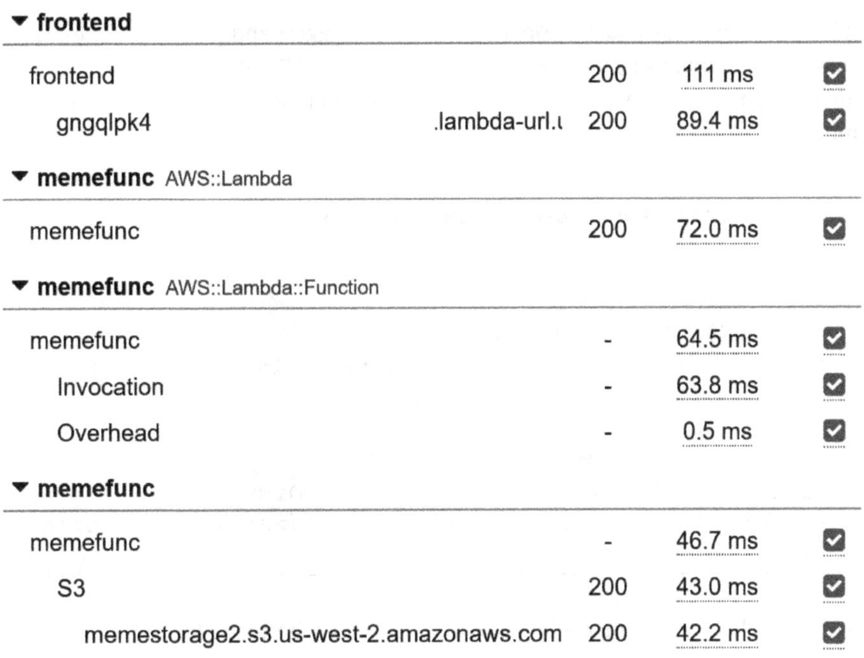

▼ **frontend**				
frontend		200	111 ms	☑
gngqlpk4	.lambda-url.t	200	89.4 ms	☑
▼ **memefunc** AWS::Lambda				
memefunc		200	72.0 ms	☑
▼ **memefunc** AWS::Lambda::Function				
memefunc		-	64.5 ms	☑
Invocation		-	63.8 ms	☑
Overhead		-	0.5 ms	☑
▼ **memefunc**				
memefunc		-	46.7 ms	☑
S3		200	43.0 ms	☑
memestorage2.s3.us-west-2.amazonaws.com		200	42.2 ms	☑

Figure 3.11 – Lambda tracing with OpenTelemetry

If you check the service map, it now shows S3 in addition to Lambda nodes.

Now that you know how to enable tracing for AWS Lambda, let's see what Azure Functions are capable of.

Azure Functions

Azure Functions support distributed tracing with Azure Monitor (Application Insights) out-of-the-box. It includes triggers and most bindings. If you use in-process functions, tracing covers user code too, with isolated workers, you need to enable and configure tracing in the worker process yourself.

Azure Functions rely on the instrumentations in client SDKs used for triggers and bindings. For example, they reuse ASP.NET Core Activities in HTTP Trigger and Azure SDK instrumentation for Azure Blob Storage inputs and outputs.

The Azure Functions runtime does not support OpenTelemetry for in-process functions yet, but your observability vendor may provide an extension that covers this gap.

In our sample, Azure Functions host automatically reports triggers and binding calls to Application Insights – this auto-collection lights up in presence of the APPLICATIONINSIGHTS_CONNECTION_STRING environment variable, which we can set in the local.settings.json file, as shown in this example:

./serverless/azure/memefunc/local.settings.json

```
"Values": {
    ...
    "APPLICATIONINSIGHTS_CONNECTION_STRING":
        "InstrumentationKey=<key>;IngestionEndpoint=
            <endpoint>"
}
```

https://github.com/PacktPublishing/Modern-Distributed-Tracing-in-.NET/blob/main/chapter3/serverless/azure/memefunc/local.settings.json

We also need to enable OpenTelemetry for the worker process with the following code:

./serverless/azure/memefunc/Program.cs

```
var host = new HostBuilder()
   .ConfigureFunctionsWorkerDefaults
   .ConfigureServices(services => services
     .AddOpenTelemetry()
     .WithTracing(builder => builder
       .AddSource("Microsoft.Azure.Functions.Worker")
       ...)
     )
   .Build();
```

https://github.com/PacktPublishing/Modern-Distributed-Tracing-in-.NET/blob/main/chapter3/serverless/azure/memefunc/Program.cs

Here we use a familiar way to enable OpenTelemetry, but the Microsoft.Azure.Functions.Worker activity source is new. The source is part of Azure Functions Worker and propagates trace context from the host to isolated worker. It creates an activity representing worker invocation.

On the **frontend** side, we use Azure.Monitor.OpenTelemetry.Exporter to send data to Application Insights endpoint directly.

To run the sample, we'll need an Application Insights resource. You can create one with the following command:

```
$ az monitor app-insights component create --app <resource-
  name> --location <region> -g <resource-group>
```

It will return JSON output containing connectionString, which we'll need to configure Functions. Let's now set Azure Blob Storage and Application Insights connection strings in memefunc/local. setting.json and we're ready to run the application:

```
serverless/azure/frontend$ dotnet run
serverless/azure/memefunc$ func start --port 5050
```

Hit **frontend** at http://localhost:5051 to upload and download some memes, and then go to your Application Insights resource and search for recent requests. *Figure 3.12* shows an example of captured trace:

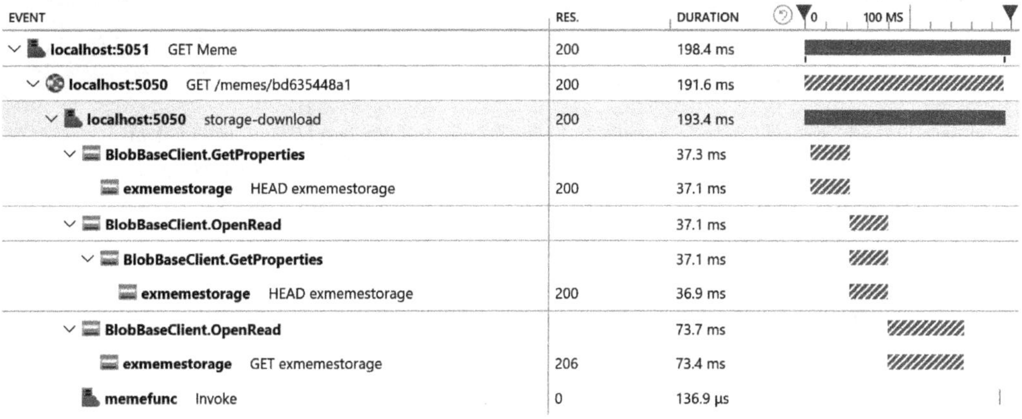

EVENT	RES.	DURATION	
∨ 🖥 **localhost:5051** GET Meme	200	198.4 ms	
∨ 🌐 **localhost:5050** GET /memes/bd635448a1	200	191.6 ms	
∨ 🖥 **localhost:5050** storage-download	200	193.4 ms	
∨ ▦ **BlobBaseClient.GetProperties**		37.3 ms	
▦ **exmemestorage** HEAD exmemestorage	200	37.1 ms	
∨ ▦ **BlobBaseClient.OpenRead**		37.1 ms	
∨ ▦ **BlobBaseClient.GetProperties**		37.1 ms	
▦ **exmemestorage** HEAD exmemestorage	200	36.9 ms	
∨ ▦ **BlobBaseClient.OpenRead**		73.7 ms	
▦ **exmemestorage** GET exmemestorage	206	73.4 ms	
🖥 **memefunc** Invoke	0	136.9 µs	

Figure 3.12 – Azure Functions trace

We traced this call from **frontend** and into the storage-download function that in turn downloaded a blob. We used Azure Blob Storage bindings, so all the communication with blob storage was handled by Azure Functions host and outside of the worker process. As a result, the Azure Functions invocation span (storage-download) and all spans related to blobs are reported by the Functions host.

The Invoke span is recorded by Microsoft.Azure.Functions.Worker activity source; it represents function invocation on the worker side. If we had any nested operations done inside worker, we'd see them reported as children of the Invoke span.

Even though most of the application logic happened outside of the application code, we can see what happened under the hood because of tracing.

Summary

In this chapter, we explored instrumentations in the .NET ecosystem. You learned how to evaluate and configure different kinds of instrumentation libraries, how to enable and use tracing on Dapr, and what serverless environments can provide with different levels of configuration.

Client library auto-instrumentations can be found in OpenTelemetry repositories or registries, while some libraries don't need instrumentations, providing tracing natively. Instrumentations' maturity and stability levels vary, so it's important to review and test them as a part of your normal integration and stress testing. Instrumentations usually provide configuration options to control the amount of details they capture, allowing you to find the right cost-value ratio for your system. Client libraries and frameworks are not the only sources of traces—your infrastructure such as service meshes, web servers, load balancers, and proxies can emit them. We checked out the tracing story in Dapr and confirmed that it provides insights into Dapr itself but can't propagate the context and stamp it on the logs and other signals in the application. So, infrastructure traces complement but cannot substitute in-process tracing.

Serverless environments provide integration with tracing and monitoring tools; it's critical for them since users are limited in the configuration of serverless runtime.

We explored AWS Lambda, which supports OpenTelemetry, with ADOT Collector and in-code configuration, and Azure Functions that supports vendor-specific codeless instrumentation for in-process mode, while out-of-the-box OpenTelemetry support is yet to come.

Now that you know how to discover and use third-party instrumentations in different environments, you should be able to get observability into a broad spectrum of distributed applications. However, to debug in-process issues such as deadlocks, memory leaks, or inefficient code, we'll need lower-level telemetry—this is what we're going to explore in the next chapter.

Questions

1. How would you find instrumentation for a popular library you use? When you find one, what would you check for?

2. What is a typical mechanism behind OpenTelemetry tracing instrumentations?

3. What service mesh can and cannot do in terms of tracing?

4

Low-Level Performance Analysis with Diagnostic Tools

While distributed tracing works great for microservices, it's less useful for deep performance analysis within a process. In this chapter, we'll explore .NET diagnostics tools that allow us to detect and debug performance issues and profile inefficient code. We'll also learn how to perform ad hoc performance analysis and capture necessary information automatically in production.

In this chapter, you will learn how to do the following:

- Use .NET runtime counters to identify common performance problems
- Use performance tracing to optimize inefficient code
- Collect diagnostics in production

By the end of this chapter, you will be able to debug memory leaks, identify thread pool starvation, and collect and analyze detailed performance traces with .NET diagnostics tools.

Technical requirements

The code for this chapter is available in this book's repository on GitHub at `https://github.com/PacktPublishing/Modern-Distributed-Tracing-in-.NET/tree/main/chapter4`. It consists of the following components:

- The `issues` application, which contains examples of performance issues
- `loadgenerator`, which is a tool that generates load to reproduce problems

To run samples and perform analysis, we'll need the following tools:

- .NET SDK 7.0 or later.

- The .NET `dotnet-trace`, `dotnet-stack`, and `dotnet-dump` diagnostics tools. Please install each of them with `dotnet tool install –global dotnet-<tool>`.

- Docker and `docker-compose`.

To run the samples in this chapter, start the observability stack, which consists of Jaeger, Prometheus, and the OpenTelemetry collector, with `docker-compose up`.

Make sure that you start the **issues** app in the Release configuration – for example, by calling `dotnet run -c Release` from the `issues` folder. We don't run it in Docker so that it's easier to use diagnostics tools.

In the **issues** app, we enabled the .NET runtime and process counters with the help of the `OpenTelemetry.Instrumentation.Process` and `OpenTelemetry.Instrumentation.Runtime` NuGet packages and configured metrics for the HTTP client and ASP.NET Core. Here's our metrics configuration:

Program.cs

```
builder.Services.AddOpenTelemetry()
    ...
    .WithMetrics(meterProviderBuilder =>
        meterProviderBuilder
        .AddOtlpExporter()
            .AddProcessInstrumentation()
            .AddRuntimeInstrumentation()
            .AddHttpClientInstrumentation()
            .AddAspNetCoreInstrumentation());
```

https://github.com/PacktPublishing/Modern-Distributed-Trac-ing-in-.NET/tree/main/chapter4/issues/Program.cs

---

Investigating common performance problems

Performance degradation is a symptom of some other issues such as race conditions, dependency slow-down, high load, or any other problem that causes your **service-level indicators** (**SLIs**) to go beyond healthy limits and miss **service-level objectives** (**SLOs**). Such issues may affect multiple, if not all, code paths and APIs, even if they're initially limited to a specific scenario.

For example, when a downstream service experiences issues, it can cause throughput to drop significantly for all APIs, including those that don't depend on that downstream service. Retries, additional connections, or threads that handle downstream calls consume more resources than usual and take them away from other requests.

> **Note**
>
> Resource consumption alone, be it high or low, does not indicate a performance issue (or lack of it). High CPU or memory utilization can be valid if users are not affected. It could still be important to investigate when they are unusually high as it could be an early signal of a problem to come.

We can detect performance issues by monitoring SLIs and alerting them to violations. If you see that issues are widespread and not specific to certain scenarios, it makes sense to check the overall resource consumption for the process, such as CPU usage, memory, and thread counts, to find the bottleneck. Then, depending on the constrained resource, we may need to capture more information, such as dumps, thread stacks, detailed runtime, or library events. Let's go through several examples of common issues and talk about their symptoms.

Memory leaks

Memory leaks happen when an application consumes more and more memory over time. For example, if we cache objects in-memory without proper expiration and overflow logic, the application will consume more and more memory over time. Growing memory consumption triggers garbage collection, but the cache keeps references to all objects and GC cannot free them up.

Let's reproduce a memory leak and go through the signals that would help us identify it and find the root cause. First, we need to run the **issues** app and then add some load using the `loadgenerator` tool:

```
loadgenerator$ dotnet run -c Release memory-leak --parallel 100
--count 20000000
```

It makes 20 million requests and then stops, but if we let it run for a long time, we'll see throughput dropping, as shown in *Figure 4.1*:

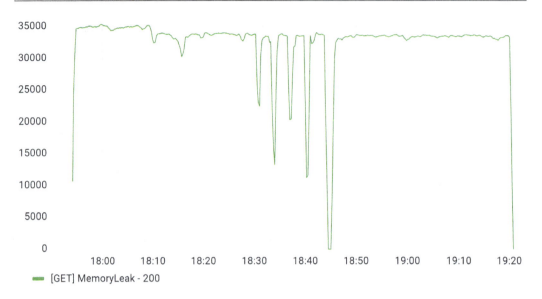

Figure 4.1 – Service throughput (successful requests per second)

We can see periods when throughput drops and the service stops processing requests – let's investigate why.

.NET reports event counters that help us monitor the size of each GC generation. Newly allocated objects appear in **generation 0**; if they survive garbage collection, they get promoted to **generation 1**, and then to **generation 2**, where they stay until they're collected or the process terminates. Large objects (that are 85 KB or bigger) appear on a **large object heap** (**LOH**).

OpenTelemetry runtime instrumentations report generation sizes under the process_runtime_dotnet_gc_heap_size_bytes metric. It's also useful to monitor the **physical** memory usage reported by OpenTelemetry process instrumentation as process_memory_usage_bytes. We can see generation 2 and physical memory consumption in *Figure 4.2*:

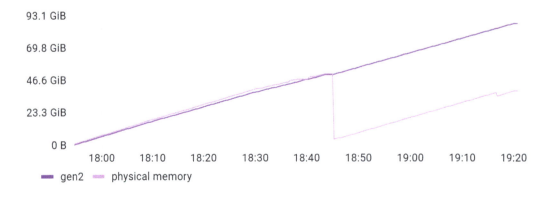

Figure 4.2 – Memory consumption showing a memory leak in the application

We can see that generation 2 grows over time, along with the virtual memory. The physical memory used by the process goes up and down, which means that the OS started using disk in addition to RAM. This process is called **paging** or **swapping**, which is enabled (or disabled) at the OS level. When enabled, it may significantly affect performance since RAM is usually much faster than disk.

Eventually, the system will run out of physical memory and the pagefile will reach its size limit; then, the process will crash with an `OutOfMemoryException` error. This may happen earlier, depending on the environment and heap size configuration. For 32-bit processes, OOM happens when the virtual memory size reaches 4 GB as it runs out of address space. Memory limits can be configured or imposed by the application server (IIS), hosting providers, or container runtimes.

Kubernetes or Docker allows you to limit the virtual memory for a container. The behavior of different environments varies, but in general, the application is terminated with the `OutOfMemory` exit code after the limit is reached. It might take days, weeks, or even months for a memory leak to crash the process with `OutOfMemoryException`, so some memory leaks can stay dormant, potentially causing rare restarts and affecting only a long tail of latency distribution.

Memory leaks on the hot path can take the whole service down fast. When memory consumption grows quickly, garbage collection intensively tries to free up some memory, which uses the CPU and can pause managed threads.

We can monitor garbage collection for individual generations using .NET event counters and OpenTelemetry instrumentation, as shown in *Figure 4.3*:

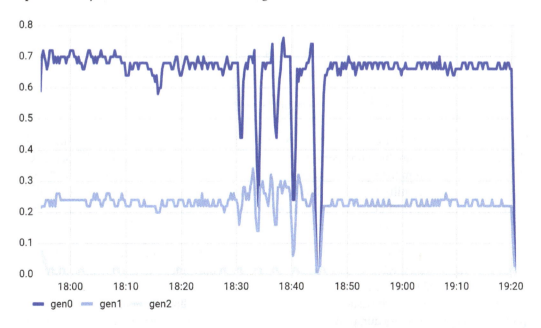

Figure 4.3 – Garbage collection rate per second for individual generations

As you can see, the generation 0 and generation 1 collections happened frequently. Looking at the consistent memory growth and the frequency of garbage collection, we can now be pretty sure we're dealing with a memory leak. We could also collect GC events from the `Microsoft-Windows-DotNETRuntime` event provider (we'll learn how to do this in the next section) to come to the same conclusion.

Let's also check the CPU utilization (shown in *Figure 4.4*) reported by the OpenTelemetry process instrumentation as the `process_cpu_time_seconds_total` metric, from which we can derive the utilization:

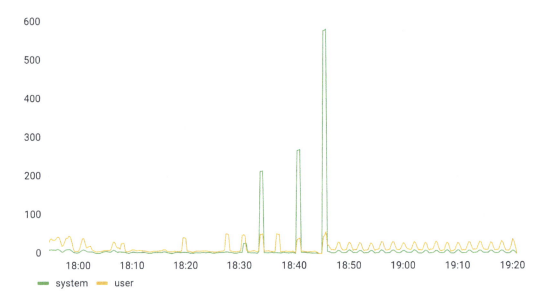

Figure 4.4 – CPU utilization during the memory leak

We can see that there are periods when both user CPU utilization and privileged (system) CPU utilization go up. These are the same periods when throughput dropped in *Figure 4.1*. User CPU utilization is derived from the `System.Diagnostics.Process.UserProcessorTime` property, while system utilization (based on OpenTelemetry terminology) is derived from the `System.Diagnostics.Process.PriviledgedProcessorTime` property. These are the same periods when throughput dropped in *Figure 4.1*.

Our investigation could have started with high latency, high error rate, a high number of process restarts, high CPU, or high memory utilization, and all of those are symptoms of the same problem – a memory leak. So, now, we need to investigate it further – let's collect a memory dump to see what's in there. Assuming you can reproduce the issue on a local machine, Visual Studio or JetBrains dotMemory can capture and analyze a memory dump. We will use `dotnet-dump`, which we can run on an instance experiencing problems. Check out the .NET documentation at `https://learn.microsoft.com/dotnet/core/diagnostics/dotnet-dump` to learn more about the tool.

So, let's capture the dump using the following command:

```
$ dotnet-dump collect --name issues
```

Once the dump has been collected, we can analyze it with Visual Studio, JetBrains dotMemory, or other tools that automate and simplify it. We're going to do this the hard way with the dotnet-dump CLI tool:

```
$ dotnet-dump analyze <dump file name>
```

This will open a prompt where we can run **SOS** commands. SOS is a debugger extension that allows us to examine running processes and dumps. It can help us find out what's on the heap.

We can do this with the dumpheap -stat command, which prints the count and total count and size of objects by their type, as shown in *Figure 4.5*:

```
00007ffe53fd62c0        34    318767920 System.Collections.Concurrent.ConcurrentQueueSegment`1+Slot[[System.Action, Syste
m.Private.CoreLib]][]
00007ffe5429a6a8 19747176     631909632 Microsoft.AspNetCore.Mvc.Infrastructure.CopyOnWriteList`1[[Microsoft.AspNetCore.M
vc.ModelBinding.IValueProviderFactory, Microsoft.AspNetCore.Mvc.Abstractions]]
00007ffe53f88600 19747176     789887040 Microsoft.AspNetCore.Routing.RouteData
00007ffe53f5f4b8 19747176     789887040 issues.Controllers.MemoryLeakController+<>c__DisplayClass3_0
00007ffe542717b0 19747181     789887240 Microsoft.AspNetCore.Routing.RouteValueDictionary
00007ffe53f53988 19747176    1105841856 Microsoft.AspNetCore.Mvc.ControllerContext
00007ffe53e56160 19747176    1105841856 Microsoft.AspNetCore.Mvc.ModelBinding.ModelStateDictionary
00007ffe53d216a0 19747192    1105842912 System.Collections.Generic.KeyValuePair`2[[System.String, System.Private.CoreLib]
,[System.Object, System.Private.CoreLib]][]
00007ffe53aa3848 19747204    1263821056 System.Action
00007ffe5429af10 19747176    1579774080 Microsoft.AspNetCore.Mvc.ModelBinding.ModelStateDictionary+ModelStateNode
00007ffe53f5d488 19747176    1579774080 issues.Controllers.MemoryLeakController
Total 213745529 objects
```

Figure 4.5 – Managed heap stats showing ~20 million MemoryLeakController instances

Stats are printed in ascending order, so the objects with the biggest total size appear at the end. Here, we can see that we have almost 20 million MemoryLeakController instances, which consume about 1.5 GB of memory. The controller instance is scoped to the request, and it seems it is not collected after the request ends. Let's find **the GC roots** – objects that keep controller instances alive.

We need to find the address of any controller instance. We can do this using its method table – the first hex number in each table row. The method table stores type information for each object and is an internal CLR implementation detail.

We can find the object address for it using another SOS command:

```
$ dumpheap -mt 00007ffe53f5d488
```

This will print a table that contains the addresses of all MemoryLeakController instances. Let's copy one of them so that we can find the GC root with it:

```
$ gcroot -all <controller-instance-address>
```

Figure 4.6 shows the path from the GC root to the controller instance printed by the gcroot command:

```
-> 0000021AE7B9E490 issues.ProcessingQueue
-> 0000021AE7B9E4B8 System.Collections.Concurrent.ConcurrentQueue`1[[System.Action, System.Private.CoreLib]]
-> 0000021AE7B9E4F8 System.Collections.Concurrent.ConcurrentQueueSegment`1[[System.Action, System.Private.CoreLib]]
-> 0000021EB7D48310 System.Collections.Concurrent.ConcurrentQueueSegment`1+Slot[[System.Action, System.Private.CoreL
ib]][]
-> 000002182A5F67E8 System.Action
-> 000002182A5F67C0 issues.Controllers.MemoryLeakController+<>c__DisplayClass3_0
-> 000002182A5F6770 issues.Controllers.MemoryLeakController
```

Figure 4.6 – ProcessingQueue is keeping the controller instances alive

We can see that issues.ProcessingQueue is holding this and other controller instances. It uses ConcurrentQueue<Action> inside. If we were to check the controller code, we'd see that we added an action that uses _logger – a controller instance variable that implicitly keeps controller instances alive:

MemoryLeakController.cs

```
_queue.Enqueue(() => _logger.LogInformation(
    "notification for {user}",
    new User("Foo", "leak@memory.net")));
```

```
https://github.com/PacktPublishing/Modern-Distributed-Trac-
ing-in-.NET/tree/main/chapter4/issues/Controllers/MemoryLeakCon-
troller.cs
```

To fix this, we'd need to stop capturing the controller's logger in action and add size limits and backpressure to the queue.

Thread pool starvation

Thread pool starvation happens when CLR does not have enough threads in the pool to process work, which can happen at startup or when the load increases significantly. Let's reproduce it and see how it manifests.

With the **issues** app running, add some load using the following commands to send 300 concurrent requests to the app:

```
$ dotnet run -c Release starve --parallel 300
```

Now, let's check what happens with the throughput and latency. You might not see any metrics or traces coming from the application or see stale metrics that were reported before the load started. If you try to hit any API on the issue application, such as http://localhost:5051/ok, it will time out.

If you check the CPU or memory for the **issues** process, you will see very low utilization – the process got stuck doing nothing. It lasts for a few minutes and then resolves – the service starts responding and reports metrics and traces as usual.

One way to understand what's going on when a process does not report metrics and traces is to use the `dotnet-counters` tool. Check out the .NET documentation at `https://learn.microsoft.com/dotnet/core/diagnostics/dotnet-counters` to learn more about the tool. Now, let's run it to see the runtime counters:

```
$ dotnet-counters monitor --name issues
```

It should print a table consisting of runtime counters that change over time, as shown in *Figure 4.7*:

```
[System.Runtime]
    % Time in GC since last GC (%)                              0
    Allocation Rate (B / 1 sec)                           155,192
    CPU Usage (%)                                               0
    Exception Count (Count / 1 sec)                             0
    GC Committed Bytes (MB)                                   605
    GC Fragmentation (%)                                   72.754
    GC Heap Size (MB)                                         297
    Gen 0 GC Count (Count / 1 sec)                              0
    Gen 0 Size (B)                                     1.0754e+08
    Gen 1 GC Count (Count / 1 sec)                              0
    Gen 1 Size (B)                                      6,629,400
    Gen 2 GC Count (Count / 1 sec)                              0
    Gen 2 Size (B)                                     39,515,408
    IL Bytes Jitted (B)                                   685,304
    LOH Size (B)                                       10,523,240
    Monitor Lock Contention Count (Count / 1 sec)              0
    Number of Active Timers                                   411
    Number of Assemblies Loaded                               140
    Number of Methods Jitted                                9,104
    POH (Pinned Object Heap) Size (B)                   7,405,328
    ThreadPool Completed Work Item Count (Count / 1 sec)        0
    ThreadPool Queue Length                                1,212
    ThreadPool Thread Count                                   411
    Time spent in JIT (ms / 1 sec)                              0
    Working Set (MB)                                          743
```

Figure 4.7 – The dotnet-counters output dynamically showing runtime counters

Here, we're interested in thread pool counters. We can see 1,212 work items waiting in the thread pool queue length and that it keeps growing along with the thread count. Only a few (if any) work items are completed per second.

The root cause of this behavior is the following code in the controller, which blocks the thread pool threads:

```
_httpClient.GetAsync("/dummy/?delay=100", token).Wait();
```

So, instead of switching to another work item, the threads sit and wait for the dummy call to complete. It affects all tasks, including those that export telemetry data to the collector – they are waiting in the same queue.

The runtime increases the thread pool size gradually and eventually, it becomes high enough to clean up the work item queue. Check out *Figure 4.8* to see thread pool counter dynamics:

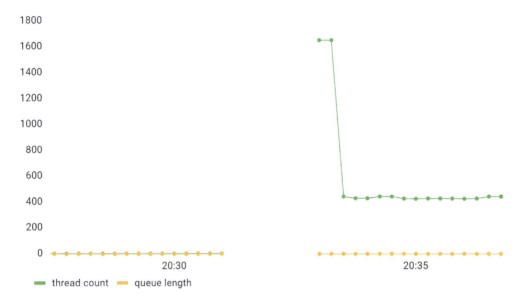

Figure 4.8 – The thread pool threads count and queue changes before and after starvation

As you can see, we have no data for the time when starvation happened. But after the thread pool queue is cleared, we start getting the data and see that the runtime adjusts the number of threads to a lower value.

We just saw how problems on a certain code path can affect the performance of the whole process and how we can use runtime metrics and diagnostics tools to narrow them down. Now, let's learn how to investigate performance issues specific to certain APIs or individual requests.

Profiling

If we analyze individual traces corresponding to thread pool starvation or memory leaks, we will not see anything special. They are fast under a small load and get slower or fail when the load increases.

However, some performance issues only affect certain scenarios, at least under typical load. Locks and inefficient code are examples of such operations.

We rarely instrument local operations with distributed tracing under the assumption that local calls are fast and exceptions have enough information for us to investigate failures.

But what happens when we have compute-heavy or just inefficient code in the service? If we look at distributed traces, we'll see high latency and gaps between spans, but we wouldn't know why it happens.

We know ahead of time that some operations, such as complex algorithms or I/O, can take a long time to complete or fail, so we can deliberately instrument them with tracing or just write a log record. But we rarely introduce inefficient code to the hot path intentionally; due to this, our ability to debug it with distributed tracing, metrics, or logs is limited.

In such cases, we need more precise signals, such as profiling. **Profiling** involves collecting call stacks, memory allocations, timings, and the frequency of calls. This can be done in-process using .NET profiling APIs that need the application to be configured in a certain way. Low-level performance profiling is usually done locally on a developer machine, but it used to be a popular mechanism among **Application Performance Monitoring** (**APM**) tools to collect performance data and traces.

In this chapter, we're going to use a different kind of profiling, also called performance tracing, which relies on `System.Diagnostics.Tracing.EventSource`, and can be done ad hoc. `EventSource` is essentially a platform logger – CLR, libraries, and frameworks write their diagnostics to event sources, which are disabled by default, but it's possible to enable and control them dynamically.

The .NET runtime and libraries events cover GC, tasks, the thread pool, the DNS, sockets, and HTTP, among other things. ASP.NET Core, Kestrel, Dependency Injection, Logging, and other libraries have their own event providers too.

You can listen to any provider inside the process using `EventListener` and access events and their payloads, but the true power of `EventSource` is that you can control providers from out-of-process over **EventPipe** – the runtime component that allows us to communicate with .NET applications. We saw it in action when we gathered event counters and collected verbose logs with the `dotnet-monitor` tool in *Chapter 2, Native Monitoring in .NET*.

Let's see how performance tracing and profiling with `EventSource` can help us investigate specific issues.

Inefficient code

Let's run our demo application and see how inefficient code can manifest itself. Make sure the observability stack is running, then start the **issues** application, and then apply some load:

```
$ dotnet run -c Release spin --parallel 100
```

The load generator bombards the `http://localhost:5051/spin?fib=<n>` endpoint with 100 concurrent requests. The spin endpoint calculates an *n*th Fibonacci number; as you'll see, our Fibonacci implementation is quite inefficient.

Assuming we don't know how bad this Fibonacci implementation is, let's try to investigate why this request takes so long. Let's open Jaeger by going to `http://localhost:16686`, clicking on **Find traces**, and checking out the latency distribution, as shown in *Figure 4.9*:

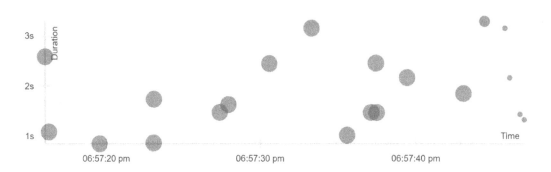

Figure 4.9 – Latency distribution in Jaeger

We can see that almost all requests take more than 2 seconds to complete. If you click on any of the dots, Jaeger will show the corresponding trace. It should look similar to the one shown in *Figure 4.10*:

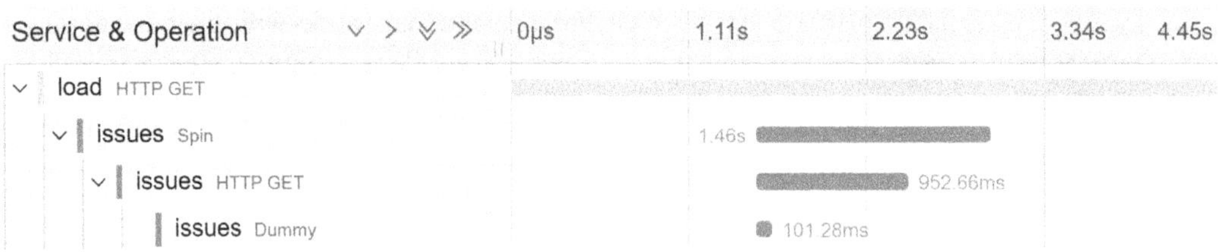

Figure 4.10 – Long trace in Jaeger

The load application is instrumented so that we can measure client latency too. We can see that the client request took 4.5 seconds, while the server-side request took about 1.5 seconds. In a spin request, we call the dummy controller of the same application and can see corresponding client and server spans. The only thing that stands out here is that there are plenty of gaps and we don't know what happened there.

If we check out the metrics, we will see high CPU and high server latency, but nothing suspicious that can help us find the root cause. So, it's time to capture some performance traces.

Multiple tools can capture performance traces for the process that experiences this issue, such as PerfView on Windows, or PerfCollect on Linux.

We're going to use the cross-platform dotnet-trace CLI tool, which you can install and use anywhere. Go ahead and run it using the following command for 10-20 seconds:

```
$ dotnet-trace collect --name issues
```

With this command, we've enabled the Microsoft-DotNETCore-SampleProfiler event source (among other default providers) to capture managed thread call stacks for the **issues** application. You can find out more about the dotnet-trace tool by reading the .NET documentation at https://learn.microsoft.com/dotnet/core/diagnostics/dotnet-trace. We could also configure it to collect events from any other event source.

The tool saves traces to the `issues.exe_*.nettrace` file, which we can analyze with it as well:

```
$ dotnet-trace report issues.exe_*.nettrace topN
```

It outputs the top (5 by default) methods that have been on the stack most of the time. *Figure 4.11* shows some sample output:

```
Top 5 Functions (Exclusive)                                         Inclusive      Exclusive
1. Missing Symbol                                                   56.47%         56.47%
2. SpinController.MostInefficientFibonacci(int32)                   31.74%         29.3%
3. Monitor.Wait(class System.Object,int32)                          2.65%          2.65%
4. WaitHandle.WaitOneNoCheck(int32)                                 2.63%          2.63%
5. WaitHandle.WaitMultiple(value class System.ReadOnlySpan`1<class System    2.6%      2.6%
```

Figure 4.11 – Top five methods on the stack

There are no details about the top line – this is due to unmanaged or dynamically generated code. But the second one is ours – the `MostInefficientFibonacci` method looks suspicious and is worth checking. It was on the call stack 29.3% of the time (exclusive percentage). Alongside nested calls, it was on the call stack 31.74% of the time (inclusive percentage). This was easy, but in more complex cases, this analysis won't be enough, and we might want to dig even further into popular call stacks.

You can open the trace file with any of the performance analysis tools I mentioned previously. We'll use SpeedScope (`https://www.speedscope.app/`), a web-based tool.

First, let's convert the trace file into `speedscope` format:

```
dotnet-trace convert --format speedscope
    issues.exe_*.nettrace
```

Then, we must drop the generated JSON file into SpeedScope via the browser. It will show the captured call stacks for each thread.

You can click through different threads. You will see that many of them are sitting and waiting for work, as shown in *Figure 4.12*:

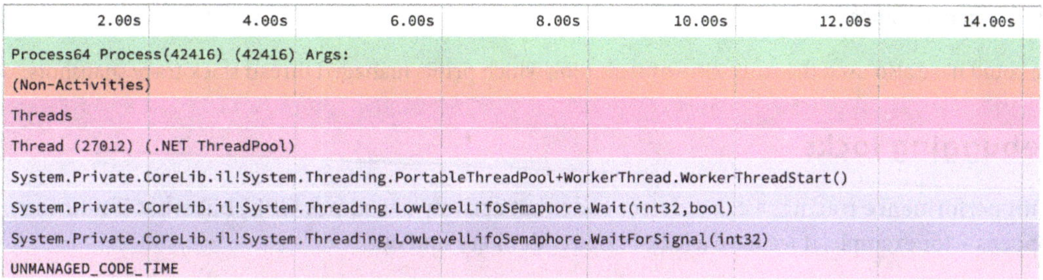

Figure 4.12 – The thread is waiting for work

This explains the top line in the report – most of the time, threads are waiting in unmanaged code.

There is another group of threads that is working hard to calculate Fibonacci numbers, as you can see in *Figure 4.13*:

Figure 4.13 – Call stack showing controller invocation with Fibonacci number calculation

As you can see, we use a recursive Fibonacci algorithm without memorization, which explains the terrible performance.

We could have also used the `dotnet-stack` tool, which prints managed thread stack trace snapshots.

Debugging locks

With performance tracing, we can detect code that actively consumes CPU, but what if nothing happens – for example, if we have a lock in our code? Let's find out.

Let's start the **issues** app and generate some load:

```
$dotnet run -c Release lock --parallel 1000
```

If we check the CPU and memory consumption, we can see that they are low and don't grow much, the thread count doesn't change much, the thread queue is empty, and the contention rate is low. At the same time, the throughput is low (around 60 requests per second) and the latency is big (P95 is around 3 seconds). So, the application is doing nothing, but it can't go faster. If we check the traces, we will see a big gap with no further data.

This issue is specific to the lock API; if we hit another API, such as http://localhost:5051/ ok, it responds immediately. This narrows down our search for the lock API.

Assuming we don't know there is a lock there, let's collect some performance traces again with $ dotnet-trace collect --name issues. If we get the topN stacks, as in the previous example, we won't see anything interesting – just threads waiting for work – locking is fast; waiting for the locked resource to become available takes much longer.

We can dig deeper into the generated trace file to find actual stack traces on what happens in the lock controller. We're going to use PerfView on Windows, but you can use PerfCollect on Linux, or other tools such as JetBrains dotTrace to open trace files and find stack traces.

Let's open the trace file with PerfView and then click on the **Thread Time** option – it will open a new window. Let's fold all the threads and search for LockController.Lock, as shown in *Figure 4.14*:

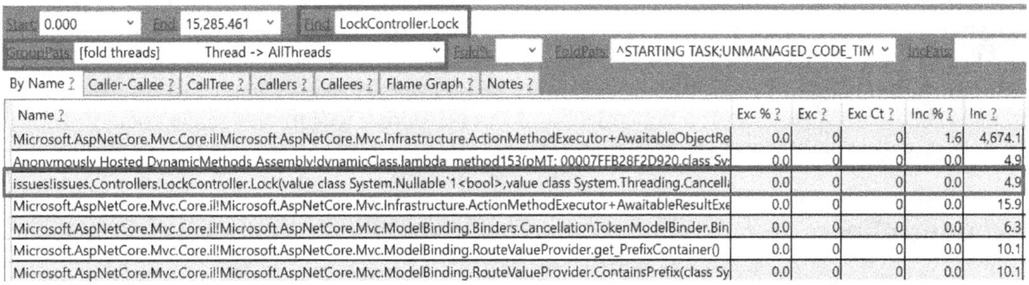

Name ?	Exc % ?	Exc ?	Exc Ct ?	Inc % ?	Inc ?
Microsoft.AspNetCore.Mvc.Core.il!Microsoft.AspNetCore.Mvc.Infrastructure.ActionMethodExecutor+AwaitableObjectRe	0.0	0	0	1.6	4,674.1
Anonymously Hosted DynamicMethods Assembly!dynamicClass.lambda_method153(pMT: 00007FFB28F2D920,class Sy	0.0	0	0	0.0	4.9
issues!issues.Controllers.LockController.Lock(value class System.Nullable`1<bool>,value class System.Threading.Cancell.	0.0	0	0	0.0	4.9
Microsoft.AspNetCore.Mvc.Core.il!Microsoft.AspNetCore.Mvc.Infrastructure.ActionMethodExecutor+AwaitableResultExe	0.0	0	0	0.0	15.9
Microsoft.AspNetCore.Mvc.Core.il!Microsoft.AspNetCore.Mvc.ModelBinding.Binders.CancellationTokenModelBinder.Bin	0.0	0	0	0.0	6.3
Microsoft.AspNetCore.Mvc.Core.il!Microsoft.AspNetCore.Mvc.ModelBinding.RouteValueProvider.get_PrefixContainer()	0.0	0	0	0.0	10.1
Microsoft.AspNetCore.Mvc.Core.il!Microsoft.AspNetCore.Mvc.ModelBinding.RouteValueProvider.ContainsPrefix(class Sy	0.0	0	0	0.0	10.1

Figure 4.14 – Finding LockController stacks across all threads

We can see that LockController rarely appears on the call stack, as well as its nested calls – we can tell since both the inclusive and exclusive percentages are close to 0. From this, we can conclude that whatever we're waiting for is asynchronous; otherwise, we would see it on the call stack.

Now, let's right-click on the LockController line and click on **Drill-Down**. It will open yet another window focused on LockController stacks. Switch to the **CallTree** tab, as shown in *Figure 4.15*:

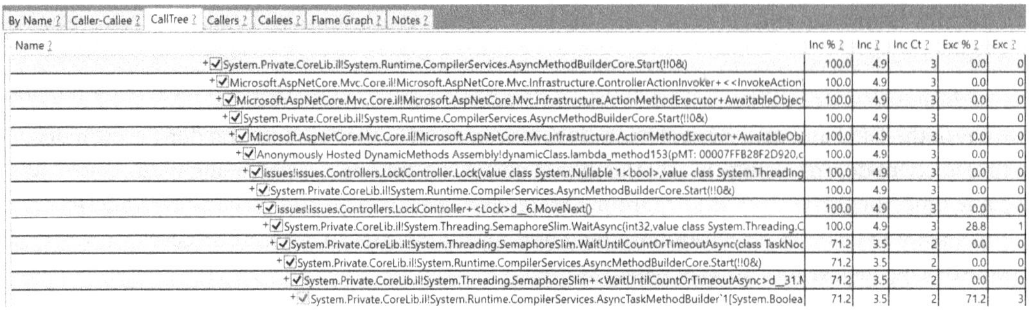

Figure 4.15 – Call stack with LockController.Lock

We can see that the controller calls `SemaphoreSlim.WaitAsync` – this should be our first suspect. It would explain the low CPU, low memory usage, and no anomalies in the thread counts. It still makes clients wait and keeps client connections open.

> **Note**
>
> We can only see the synchronous part of the call stack in *Figure 4.15* – it does not include `WaitAsync` or anything that happens after that.

The analysis we've done here relies on luck. In real-world scenarios, this issue would be hidden among other calls. We would have multiple suspects and would need to collect more data to investigate further. Since we're looking for asynchronous suspects, collecting task-related events with `dotnet-trace` from the `System.Threading.Tasks.TplEventSource` provider would be useful.

The issue is obvious if we look into the code, but it can be hidden well in real-world code, behind feature flags or third-party libraries:

LockController.cs

```
await semaphoreSlim.WaitAsync(token);
try
{
    ThreadUnsafeOperation();
    await _httpClient.GetAsync("/dummy/?delay=10", token);
}
finally
{
    semaphoreSlim.Release();
}
```

```
https://github.com/PacktPublishing/Modern-Distributed-Trac-
ing-in-.NET/tree/main/chapter4/issues/Controllers/LockController.
cs
```

The problem here is that we put a lock around the HTTP call to the downstream service. If we wrap only `ThreadUnsafeOperation` into a synchronous lock, we'll see a much higher throughput of around 20K requests per second and low latency with P95 of around 20 milliseconds.

Performance tracing is a powerful tool that allows us to capture low-level data reported by the .NET runtime, standard, and third-party libraries. In the examples we have covered in this chapter, we run diagnostics tools ad hoc and on the same host as the service. This is reasonable when you're reproducing issues locally or optimizing your service on the dev box. Let's see what we can do in a more realistic case with multiple instances of services running and restricted SSH access.

Using diagnostics tools in production

In production, we need to be able to collect some data proactively with reasonable performance and a telemetry budget so that we can analyze data afterward.

It's difficult to reproduce an issue on a specific instance of a running process and collect performance traces or dumps from it in a secure and distributed application. If an issue such as a slow memory leak or a rare deadlock affects just a few instances, it might be difficult to even detect it and, when detected, the instance has already been recycled and the issue is no longer visible.

Continuous profiling

What we're looking for is a continuous profiler – a tool that collects sampled performance traces. It can run for short periods to minimize the performance impact of collection on each instance and send profiles to central storage, where they can be stored, correlated with distributed traces, queried, and viewed. Distributed tracing supports sampling and a profiler can use it to capture traces and profiles consistently.

Many observability vendors, such as Azure Monitor, New Relic, Dynatrace, and others, provide continuous profilers for .NET. For example, Azure Monitor allows us to navigate to profiles from traces, as you can see in *Figure 4.16*:

Figure 4.16 – Navigating to a profile from a trace in Azure Monitor

We will see a long trace for the inefficient code examples we went through earlier in this chapter, but the continuous profiler was enabled and captured some of these calls. If we click on the profiler icon, we will see the call stack, similar to the one we captured with `dotnet-collect`, as shown in *Figure 4.17*:

EVENTS	MODULE	THREAD TIME ↓	TIMELINE
▼ 🔥 Request(/id=-1)		1674.68 ms	
▼ 🔥 Framework/Library Thread.StartCallback	System.Private.CoreLib.il ⊗	1674.68 ms	
▼ 🔥 dynamicClass.lambda_method132	anonymously hosted dynamicmethods assembly ⊗	1672.93 ms	
🔥 SpinController.Spin	issues ⊗	1672.93 ms	
🔥 Framework/Library AsyncMethodBuilderCore.Start	System.Private.CoreLib.il ⊗	1672.93 ms	
▼ 🔥 SpinController+<Spin>d__2.MoveNext	issues ⊗	1672.93 ms	
▼ 🔥 Framework/Library AsyncTaskMethodBuilder`1.AwaitUnsafeOnCompleted	System.Private.CoreLib.il ⊗	1559.55 ms	
▼ 🔥 SpinController+<Spin>d__2.MoveNext	issues ⊗	1558.60 ms	
🔥 SpinController.MostInefficientFibonachi	issues ⊗	1558.60 ms	
🔥 SpinController.MostInefficientFibonachi	issues ⊗	1558.60 ms	
🔥 SpinController.MostInefficientFibonachi	issues ⊗	1558.60 ms	
🔥 SpinController.MostInefficientFibonachi	issues ⊗	1558.60 ms	
▶ 🔥 SpinController.MostInefficientFibonachi	issues ⊗	1558.60 ms	
CPU Time		0.95 ms	
▶ 🖳 Framework/Library AsyncMethodBuilderCore.Start	System.Private.CoreLib.il ⊗	113.38 ms	

Figure 4.17 – Profile showing a recursive call stack with the MostInefficientFibonacci method

With a continuous profiler, we can debug inefficient code in a matter of seconds, assuming that the problem is reproduced frequently enough so that we can capture both distributed trace and profile for it.

The dotnet-monitor tool

Beyond profiling individual calls, we also need to be able to capture dumps proactively and on demand. It's possible to configure .NET to capture dumps when a process crashes, but it doesn't always work in containers and it's not trivial to access and transfer dumps.

With `dotnet-monitor`, we can capture logs, memory, and GC dumps, and collect performance traces in the same way we did with `dotnet` diagnostic tools:

- Performance traces from event sources can be collected with the `dotnet-monitor /trace` API or the `dotnet-trace` CLI tool
- Dumps can be collected with the `/dump` API or the `dotnet-dump` tool
- Event counters can be collected with the `/metrics` API or the `dotnet-counters` tool

Check out the `dotnet-monitor` documentation to learn more about these and other HTTP APIs it provides: `https://github.com/dotnet/dotnet-monitor/tree/main/documentation`.

We can also configure triggers and rules that proactively collect traces or dumps based on CPU or memory utilization, GC frequency, and other runtime counter values. Results are uploaded to configurable external storage.

We looked at some features of `dotnet-monitor` in *Chapter 2, Native Monitoring in .NET*, where we run it as a sidecar container in Docker. Similarly, you can run it as a sidecar in Kubernetes.

Summary

Performance issues affect the user experience by decreasing service availability. Distributed tracing and common metrics allow you to narrow down the problem to a specific service, instance, API, or another combination of factors. When it's not enough, you could increase resolution by adding more spans, but at some point, the performance impact and cost of the solution would become unreasonable.

.NET runtime metrics provide insights into CLR, ASP.NET Core, Kestrel, and other components. Such metrics can be collected with OpenTelemetry, `dotnet-counters`, or `dotnet-monitor`. It could be enough to root cause an issue, or just provide input on how to continue the investigation. The next step could be capturing process dumps and analyzing memory or threads' call stacks, which can be achieved with `dotnet-dump`.

For problems specific to certain scenarios, performance traces provide details so that we can see what happens in the application or under the hood in third-party library code. Performance traces are collected with `dotnet-trace` or `dotnet-monitor`. By capturing performance traces, we can see detailed call stacks, get statistics on what consumes CPU, and monitor contentions and garbage collection more precisely. This is not only a great tool to investigate low-level issues but also to optimize your code.

Collecting low-level data in a secure, multi-instance environment is challenging. Continuous profilers can collect performance traces and other diagnostics on-demand, on some schedule, or by reacting to certain triggers. They also can take care of storing data in a central location and then visualizing and correlating it with other telemetry signals.

The `dotnet-monitor` tool can run as a sidecar and then provide essential features to diagnostics data proactively or on-demand and send it to external storage.

In this chapter, you learned how to collect diagnostics data using .NET diagnostics tools and how to use it to solve several classes of common performance issues. Applying this knowledge, along with what we learned about metrics, distributed tracing, and logs previously, should allow you to debug most distributed and local issues in your application.

So, now, you know everything you need to leverage auto-instrumentation and make use of telemetry created by someone else. In the next chapter, we'll learn how to enrich auto-generated telemetry and tailor it to our needs.

Questions

1. What would you check first to understand whether an application was healthy?

2. If you were to see a major performance issue affecting multiple different scenarios, how would you investigate it?

3. What's performance tracing and how can you leverage it?

Part 2: Instrumenting .NET Applications

This part provides an in-depth overview and practical guide for .NET tracing, metrics, logs, and beyond. We'll start by learning about OpenTelemetry configuration and then dive deep into manual instrumentation, using different signals.

This part has the following chapters:

5

Configuration and Control Plane

In the previous chapters, we learned how to enable auto-instrumentation with a few lines of code and leverage collected telemetry to debug issues and monitor performance. Auto-collected traces and metrics provide the basis for your observability solution, but they are rarely sufficient without an application context. In this chapter, we'll learn how to customize telemetry collection – enrich, adjust, or control its volume. We will dive into the following topics:

- Controlling costs with sampling
- Enriching and filtering telemetry
- Customizing context propagation
- Building a processing pipeline with the OpenTelemetry Collector

By the end of this chapter, you will be able to choose a sampling strategy and configure it in your system, efficiently enrich auto-generated traces with custom attributes, and propagate your context between services. We'll see also how to suppress noisy spans and metrics.

Technical requirements

The code for this chapter is available in the book's repository on GitHub at `https://github.com/PacktPublishing/Modern-Distributed-Tracing-in-.NET/tree/main/chapter5`, which has the following structure:

- The `sampling` application contains sampling code snippets
- The `memes` application is an improved version of the meme service from *Chapter 2, Native Monitoring in .NET*, which contains enrichment and context propagation examples

To run samples and perform analysis, we'll need the following tools:

- .NET SDK 7.0 or later

- Docker and `docker-compose`

Controlling costs with sampling

Tracing all operations gives us the ability to debug individual issues in the system, even very rare ones, but it could be impractical from a performance and telemetry storage perspective.

The performance impact of optimized and succinct instrumentation is usually low, but telemetry ingestion, processing, storage, queries, and other observability experiences could be very costly.

Observability vendors' pricing models vary – some charge for the volume of ingested traces, others for the number of events, traces, or hosts reporting data. The ingestion cost usually includes retaining telemetry for 1 to 3 months. Data retrieval and scanning are also billed by some vendors. Essentially, costs associated with sending and retrieving telemetry grow along with telemetry volume.

Realistically, we're going to be interested in a very small fraction of traces – ones that record failures, long requests, and other rare cases. We may also query a subset of traces for analytics purposes.

So, collecting all traces comes at a relatively small performance hit and could be reasonable, but storing all of them on an observability backend is rarely justified.

> **Note**
> You might consider using traces as audit logs and then need every operation to be recorded. However, we need traces to debug and resolve incidents in production, which implies a fast query time, potentially short retention, and traces being accessible by every on-call person. Audit logs usually need a different privacy and retention policy. They also don't necessarily require fast and random access.

Sampling is a technique that allows recording a subset of traces, thus reducing storage costs. There are two main approaches to sampling: **head-based** and **tail-based**.

> **Note**
> Both head-based and tail-based sampling rely on trace context propagation that needs to happen regardless of the sampling decision.

Let's take a closer look at different sampling approaches and see how and when to apply them.

Head-based sampling

With head-based sampling, the decision to record or not record a trace is done when the trace is started by the application process and is usually random (or based on information available beforehand). The assumption here is that under high-scale problems that need attention happen frequently enough to record at least some occurrences. In other words, problems that are never recorded are too rare and are likely less important. "Too rare" and "frequently enough" here totally depend on application requirements.

Head-based sampling algorithms try to be consistent so that we can capture all spans or none of them in any trace. It's achieved by either following upstream sampling decisions or making independent, but consistent, decisions on each service. Let's learn more about these approaches and also check how we can implement custom sampling solutions.

Parent-based sampling

With the **parent-based** sampling approach, the component that starts a trace makes a sampling decision and propagates it, and all downstream services follow. The decision is propagated using a sampling bit in `traceparent`. For example, as we saw in *Chapter 1, Observability Needs of Modern Applications*, `traceparent` with `00-trace1-span1-01` indicates that upstream services recorded this span, and `00-trace2-span2-00` indicates that the span was not recorded. To enable this behavior in OpenTelemetry, you can use `ParentBasedSampler`. When all services follow the parent decision, the sampling probability (or percentage or recorded traces) configured on the first component applies to all downstream services.

> **Note**
>
> With using parent-based sampling, the component that makes decision needs to be fully trusted – if it starts to record all traces, it might overload your telemetry collection pipeline and could cause your observability backend costs to skyrocket. You would normally make sampling decisions on your API gateway or frontend and would never trust sampling decisions coming from an external client.

The root component still needs to make independent sampling decision. Even though this decision can be random and other services would follow, it's a good idea to stick to OpenTelemetry (or your observability vendor) sampling algorithms and keep them consistent across the system.

Probability sampling

Another approach involves making a sampling decision on every service but doing it consistently, so `trace` that's recorded on one service would be recorded on another if they have the same sampling rate configured.

To achieve this, the **probability sampling** algorithm calculates the **sampling score** as a hash function of `trace-id`. If the score is smaller than the probability, the span is recorded, otherwise, it's dropped.

> **Note**
>
> Probability sampling reduces your costs by only recording a fraction of the traces. Unexpected loads or bursts of traffic result in prorated growth in the volume of recorded traces. OpenTelemetry for .NET does not support a fixed-rate sampler out of the box, but you can configure the collector to do it or your observability vendor may provide one.

Probability sampling is implemented in OpenTelemetry with `TraceIdRatioBasedSampler` and can be configured with the `SetSampler` method on `TracerProviderBuilder`:

Program.cs

```
builder.Services.AddOpenTelemetry()
  .WithTracing(tp => tp
    .SetSampler(new TraceIdRatioBasedSampler(0.1))
    .AddOtlpExporter());
```

https://github.com/PacktPublishing/Modern-Distributed-Tracing-in-.NET/blob/main/chapter5/sampling/Program.cs

We set the probability to `0.1` in this example, which means that 10% of all traces will be recorded.

And if we want to configure parent-based sampling on downstream components, we should set an instance of `ParentBasedSampler` instead, as shown in the following example:

Program.cs

```
tp.SetSampler(
    new ParentBasedSampler(new AlwaysOffSampler()))
```

https://github.com/PacktPublishing/Modern-Distributed-Tracing-in-.NET/blob/main/chapter5/sampling/Program.cs

We need to provide a sampler that's used when there is no parent trace context – in this example, we're sampling out all the requests that come without `traceparent`. We could customize parent-based samplers further – set samplers to handle different cases: when the parent is remote or local, and whether the parent is recorded or not.

Consistent sampling

Assuming all services have the same sampling probability configured, all spans in a trace will be recorded or dropped consistently – there will be no partial traces. However, using the same rate for all services is not always practical. You might want to configure higher sampling probability on a new service, or for one that has a small load. We can do it using a probability sampler by configuring it to different rates on different services.

As a result, we should expect that some tracers will be recorded partially. *Figure 5.1* shows an example of a partial trace:

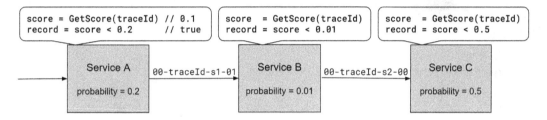

Figure 5.1 – Services with different sampling probabilities

Service A starts a trace – it generates trace-id and makes a sampling decision by calculating the score and comparing it to the sampling probability configured on the service.

Let's say the score is `0.1` – it's smaller than the probability (`0.2`), so the request is sampled in, and we should record the corresponding span and its local children. Then **Service A** calls **Service B**, which has a smaller sampling probability, `0.01`. **Service B** calculates the score – it's the same, so the decision is to not record the span or its local children. But then **Service B** calls **Service C**, which records corresponding spans.

As a result, for this trace, we'll have spans from **Service A** and **C**, but not from **Service B**. If this `trace-id` had a different score, for example, `0.005`, all services would record corresponding spans and we'd have a complete trace.

Probability sampling relies on `trace-id` being random and that the same hash function is used to calculate the score for each span. This is the case if you use vanilla OpenTelemetry in the same language on all your services and don't customize ID generation or configure vendor-specific samplers. If your `trace-id` is not random or if you have to use different sampling algorithms, we need a slightly different approach called **consistent sampling**. It's experimental and not implemented in .NET at the moment.

The approach relies on sampling score propagation: the component that starts the trace calculates the sampling score using any algorithm and propagates it to downstream services via `tracestate`. Downstream services don't need to calculate the score again – they read it from `tracestate` and make a sampling decision by comparing this score to their configured probability.

Custom sampler

You can implement your own samplers. For example, `DebugSampler` records all activities that have a `debug` flag in a `tracestate` header and uses the probability sampler for all other activities. With this sampler, you can force the trace to be recorded, by sending a request with valid `traceparent` and `tracestate: myapp=debug:1` headers, which can be useful when testing or reproducing problems:

DebugSampler.cs

```
class DebugSampler : Sampler
{
    private readonly static Sampler On
      = new AlwaysOnSampler();
    private readonly static Regex DebugFlag
      = new Regex("(^|,)myapp=debug:1($|,)",
                  RegexOptions.Compiled);

    private readonly Sampler _default;
    public DebugSampler(double probability)
    {
        _default =
            new TraceIdRatioBasedSampler(probability);
    }

    public override SamplingResult ShouldSample(
      in SamplingParameters parameters)
    {
        var tracestate =
          parameters.ParentContext.TraceState;
        if (tracestate != null &&
            DebugFlag.IsMatch(tracestate))
          return On.ShouldSample(parameters);

        return _default.ShouldSample(parameters);
    }
}
```

https://github.com/PacktPublishing/Modern-Distributed-Trac-
ing-in-.NET/blob/main/chapter5/sampling/DebugSampler.cs

The sampler implements a `ShouldSample` method, which takes `SamplingParameters` in and returns `SamplingResult`.

Sampling parameters include parent trace context and additional details such as creation-time attributes, the `Activity` name, kind, `tracestate`, and links.

`SamplingResult` is a struct that contains a `SamplingDecision` enum that takes one of three possible values:

- `Drop`: Create `Activity`, but don't record it.
- `RecordAndSample`: Create `Activity`, record it, and set the sampling flag on trace context.
- `RecordOnly`: Create `Activity` and record it, but don't set a sampling flag on trace context. Built-in samplers never return a `RecordOnly` decision, but you can implement a custom sampler and return such a decision to a trace request locally without forcing downstream services to follow it (if they respect parent decisions).

`SamplingResult` also contains updated `tracestate` values and attributes – samplers can set them, and they will be used on the activity-to-be-created. We can configure this sampler in the same way as before, by calling the `SetSampler` method on the `TracerProviderBuilder` instance.

> **Note**
>
> With OpenTelemetry, an activity is created for every sampling decision. But in .NET, it's possible to prevent a sampled-out `Activity` from being created altogether.

We'll learn more about sampling in vanilla .NET in *Chapter 6*, *Tracing Your Code*, and will see some examples of how instrumentations can suppress activity creation later in this chapter.

Tail-based sampling

With tail-based sampling, the decision is made after the trace ends and could consider trace duration, presence of errors, or any other information available on spans. As you can imagine, we should first buffer all spans in the trace and then send all of them to the observability provider or drop all of them. Tail-based sampling must happen across different services and can only be done by an external component such as the OpenTelemetry Collector. The tail-sampling processor in the Collector is highly configurable and supports multiple sampling policies, including rate-limiting, latency, and status-code-based policies. You can combine multiple policies together. It's not really possible to know when a trace ends, so the Collector starts buffering spans after the first span in that trace is received for a configurable time period and then makes a decision based on the available data.

The tail-based sampling processor in collector allows the creation of composite sampling strategies. For example, you can configure probability sampling on your .NET services to minimize performance impact and then apply rate limiting on the collector to control telemetry volume and observability backend costs.

You could also configure a higher sampling probability to collect more traces with bigger latency, errors, or specific attributes. Buffering is currently limited to a single instance of the Collector, so if spans from the same trace end up on different collectors, tail-based sampling may produce partial traces, but still would capture parts corresponding to failure or increased latency. Since tail-based sampling has to buffer spans, it is memory-consuming and requires additional compute resources, and works well for short traces. As usual, it makes sense to compare the costs of managing the Collector setup with the savings it brings.

> **Note**
>
> With non-probabilistic sampling, usage analysis based on traces is skewed and could be misleading. Metrics collected inside the application independently of traces could still be your source of truth.

Depending on your needs, you may combine different approaches – for example, collect all data but keep it in a cold(ish) storage, only using a more expensive backend for a small subset of traces.

Now you're ready to choose a sampling strategy and implement it in your system! Let's move and explore how to enrich spans with additional context.

Enriching and filtering telemetry

Traces and metrics coming from auto-instrumentation describe the technical side of an operation. While we can always add more spans with custom context (and we'll learn how to do it in *Chapter 6, Tracing Your Code*), it could be more practical to add custom context to auto-collected telemetry.

Application-specific context is necessary to track usage and contains essential information that helps to detect and investigate issues.

For example, if we take our meme service, it would be very helpful to have the meme name and size on spans. With this, we'd be able to find the most popular memes, correlate meme upload and download requests, plan capacity, make cache optimizations, or reason about partitioning.

The easiest way to add a meme name is via the `Activity.SetTag` method. For example, we can the following code on the `Meme` page:

Meme.cshtml.cs

```
public async Task<IActionResult> OnGet([FromQuery] string
  name)
{
  Activity.Current?.SetTag("meme_name", name);

    ...
}
```

https://github.com/PacktPublishing/Modern-Distributed-Tracing-in-.NET/blob/main/chapter5/memes/frontend/Pages/Meme.cshtml.cs

`Activity.Current` here is created by ASP.NET Core. It was enabled with OpenTelemetry ASP.NET Core instrumentation – if tracing was disabled, `Activity.Current` would be `null` here, so we should always guard `Activity.Current` with a `null` check or use null coalescence.

We also check whether the activity is recorded – if the `IsAllDataRequested` flag is `true`. There is no point in recording attributes on a sampled-out activity, so it's an optimization. While it's very small in this case, it's a good practice to use it to avoid any unnecessary string allocations or prevent heavier operations needed to retrieve the attribute value.

Finally, we call `SetTag` – a method that takes the string tag name and nullable object value. We'll talk more about the Activity API and tags in *Chapter 6, Tracing Your Code*.

> **Note**
>
> A tag in the .NET Activity API is the same as the OpenTelemetry `span` attribute. Tags came from OpenTracing and were not renamed for backward-compatibility reasons. This book uses *tag* and *attribute* interchangeably.

With this approach, we can add meme names to all ASP.NET Core activities. But what about HTTP client and MySQL activities? Having meme names on them would be handy.

In the general case, it can be done with a span processor, but some instrumentations provide extensibility hooks allowing them to enrich their activities.

Let's look at each of these approaches.

Span processors

The span processor is a component of the OpenTelemetry export pipeline. When OpenTelemetry is notified about activity start or stop events, it synchronously calls the corresponding method on the processor. By implementing our own processor and adding it to the tracer provider, we can intercept all activities and add attributes from `AsyncLocal`, `ThreadLocal`, or another context available globally. We could also overwrite or remove attributes or filter out activities.

Enriching

Before we do this for the meme name, we need to decide how to pass it to the processor. Since we want meme names to be on all spans from all services, it would be a good case for baggage. Baggage, as we saw in *Chapter 1, Observability Needs of Modern Applications*, represents application-specific context propagated between services.

So, let's go ahead and add the meme name to the baggage on the frontend Meme and Upload pages:

Meme.cshtml.cs

```
Activity.Current.SetTag("meme_name", name);
Baggage.SetBaggage("meme_name", name);
```

https://github.com/PacktPublishing/Modern-Distributed-Tracing-in-.NET/blob/main/chapter5/memes/frontend/Pages/Meme.cshtml.cs

Setting Baggage does not always affect an activity that has started beforehand and is current now – it's a side effect of AsyncLocal, and we'll dig into this in *Chapter 6, Tracing Your Code*. So, we're going to keep setting meme name tags on activities on the frontend pages. Baggage uses AsyncLocal underneath, so we can now reliably use it in the processor:

MemeNameEnrichingProcessor.cs

```
class MemeNameEnrichingProcessor : BaseProcessor<Activity>
{
    public override void OnEnd(Activity activity)
    {
        var name = GetName(activity);
        if (name != null)
            activity.SetTag("meme_name", name);
    }

    private string? GetName(Activity activity)
    {
        if (Baggage.Current.GetBaggage()
            .TryGetValue("meme_name", out var name))
            return name;
        return activity.GetBaggageItem("meme_name");
    }
}
```

https://github.com/PacktPublishing/Modern-Distributed-Tracing-in-.NET/blob/main/chapter5/memes/frontend/MemeNameEnriching-Processor.cs

We're overriding the OnEnd method here – we first get the meme name from the baggage and add it as a tag to the activity. We don't need to check whether the activity is null since the processor won't be called in such a case, but we might still need to check whether it's sampled in because, as we'll see soon, sampled-out activities can still sometimes reach your processor.

> **Note**
>
> We get a name from `Baggage.Current`, but if it's not there, we also check `Activity.Baggage`. The reason is that the `Baggage` type lives in the `OpenTelemetry` namespace and can be used beyond tracing. But ASP.NET Core is not aware of it and populates `Baggage` on `Activity`. As a rule of thumb, always set `Baggage` with `Baggage.SetBaggage`, but read it from `Baggage` and `Activity`.

The last step is to register this processor on `TracerProvider` on the frontend and storage services:

Program.cs

```
Builder.Services.AddOpenTelemetry()
  .WithTracing(builder => builder
    .AddProcessor<MemeNameEnrichingProcessor>()
    …);
```

https://github.com/PacktPublishing/Modern-Distributed-Tracing-in-.NET/blob/main/chapter5/memes/frontend/Program.cs

That's it – the frontend adds a meme name to `Baggage`, then `Baggage` is automatically propagated to **storage**. Whenever any activity ends, the enriching processor stamps the meme name on it.

By the way, we can also use the processor to remove unwanted tags by calling `SetTag` with a `null` value.

Filtering

Sometimes you want to drop some activities – for example, those that represent retrieving static files on a frontend or requests from web crawlers.

Dropping an activity after it has started in the middle of a trace breaks causation. It should be only done for activities that don't have any children.

Some instrumentations provide a hook to suppress activities so they are never even created – we'll see some examples of it later in this section.

But instrumentations don't always support suppression, and filtering out already started activities might be the only option. Let's see how to do it with a processor:

StaticFilesFilteringProcessor.cs

```
public class StaticFilesFilteringProcessor :
    BaseProcessor<Activity>
{
    public override void OnEnd(Activity activity)
```

```
    {
        if (activity.Kind == ActivityKind.Server &&
            activity.GetTagItem("http.method") as string
                                          == "GET" &&
            activity.GetTagItem("http.route") == null)
            activity.ActivityTraceFlags &=
                ~ActivityTraceFlags.Recorded;
    }
}
```

https://github.com/PacktPublishing/Modern-Distributed-Trac-ing-in-.NET/blob/main/chapter5/memes/frontend/StaticFilesFilter-ingProcessor.cs

In this processor, we check whether the activity has a `Server` kind (describes an incoming request), a `GET` method, and does not have a route. We can only check route presence in the `OnEnd` callback as the route is calculated after the activity starts.

Hence, we unset the recording flag on `Activity` so it will be dropped later in the exporting pipeline.

You may come up with a better heuristic to identify static files, and if it doesn't require a route, you could suppress such activities using ASP.NET Core instrumentation options, as we'll see a bit later.

To register this processor, add it to `TracerProviderBuilder` with the `AddProcessor` method. Make sure to add processors in the order you want them to run:

Program.cs

```
Builder.Services.AddOpenTelemetry()
  .WithTracing(builder => builder
    .AddProcessor<StaticFilesFilteringProcessor>()
    .AddProcessor<MemeNameEnrichingProcessor>()
    ...
);
```

https://github.com/PacktPublishing/Modern-Distributed-Trac-ing-in-.NET/blob/main/chapter5/memes/frontend/StaticFilesFilter-ingProcessor.cs

We just learned how to filter and enrich activities with processors; let's now see what we can do with instrumentation options.

Customizing instrumentations

Instrumentations may provide configuration options allowing to customize telemetry collection. For example, you can configure recording exception events with HTTP client and ASP.NET Core instrumentations through the `RecordException` flag on the corresponding configuration options:

```
AddHttpClientInstrumentation(o => o.RecordException = true)
```

Instrumentations can also provide callbacks allowing to populate attributes from instrumentation-specific contexts such as `request` or `response` objects.

Let's use it to set a request size on incoming HTTP requests on storage, so we can analyze meme sizes using ASP.NET Core instrumentation enrichments hooks:

Program.cs

```
AddAspNetCoreInstrumentation(o =>
{
    o.EnrichWithHttpRequest = (activity, request) =>
        activity.SetTag("http.request_content_length",
                        request.ContentLength);

    o.EnrichWithHttpResponse = (activity, response) =>
        activity.SetTag("http.response_content_length",
                        response.ContentLength);

    o.RecordException = true;

})
```

https://github.com/PacktPublishing/Modern-Distributed-Tracing-in-.NET/blob/main/chapter5/memes/storage/Program.cs

In addition to ASP.NET Core, you can find similar hooks for HTTP, gRPC, and SQL client instrumentations available in the `opentelemetry-dotnet` repo.

The same instrumentations also provide hooks that prevent `Activity` from being created. For example, if we wanted to suppress activities created for static files instead of dropping them in the processor, we could write something like this:

Program.cs

```
AddAspNetCoreInstrumentation(o => o.Filter =
    ctx => !IsStaticFile(ctx.Request.Path))
```

```
...
static bool IsStaticFile(PathString requestPath)
{
    return requestPath.HasValue &&
        (requestPath.Value.EndsWith(".js") ||
         requestPath.Value.EndsWith(".css"));
}
```

https://github.com/PacktPublishing/Modern-Distributed-Trac-ing-in-.NET/blob/main/chapter5/memes/frontend/Program.cs

If you consider suppressing or enriching individual activities based on dynamic context, instrumentation hooks, when available, are the best option. If you want to enrich all activities with ambient context, processors would be the right choice. Let's now see how to populate static context on all activities with resources.

Resources

An OpenTelemetry resource describes a service instance – it's a set of static attributes describing the service name, version, namespace, instance, or any other static property. OpenTelemetry defines semantic conventions for Kubernetes, generic containers, clouds, processes, OS, devices, and other common resource kinds.

You can configure resources explicitly or through environment variables. For example, we already used the OTEL_SERVICE_NAME environment variable to configure the service name. We can set OTEL_RESOURCE_ATTRIBUTES to a list of comma-separated key-value pairs (for example, region=westus,tag=foo) to specify any custom resources.

Explicit configuration can be done with ResourceBuilder, which we should register on TraceProviderBuilder:

Program.cs

```
var env = new KeyValuePair<string, object>("env",
  builder.Environment.EnvironmentName);
var resourceBuilder = ResourceBuilder.CreateDefault()
    .AddService("frontend", "memes", "1.0.0")
    .AddAttributes(new[] { env });
...
Builder.Services.AddOpenTelemetry()
  .WithTracing(builder => builder
    .SetResourceBuilder(resourceBuilder)
    ...
  );
```

```
https://github.com/PacktPublishing/Modern-Distributed-Trac-
ing-in-.NET/blob/main/chapter5/memes/frontend/Program.cs
```

Note that environment variable detection is done by default, but you can turn it off by using `ResourceBuilder.CreateEmpty` instead of the `CreateDefault` factory method. Resources are populated on each of the signals, and we can configure different resources for traces, metrics, and logs if needed.

We're ready to run the meme service and check out the result. Go ahead and run it with `compose up --build`. After the application starts, hit the frontend at `http://locahost:5051` and upload and download some memes. Now, you should be able to see new attributes – meme names on all spans and content size on incoming requests on storage. *Figure 5.2* shows an example of a GET request:

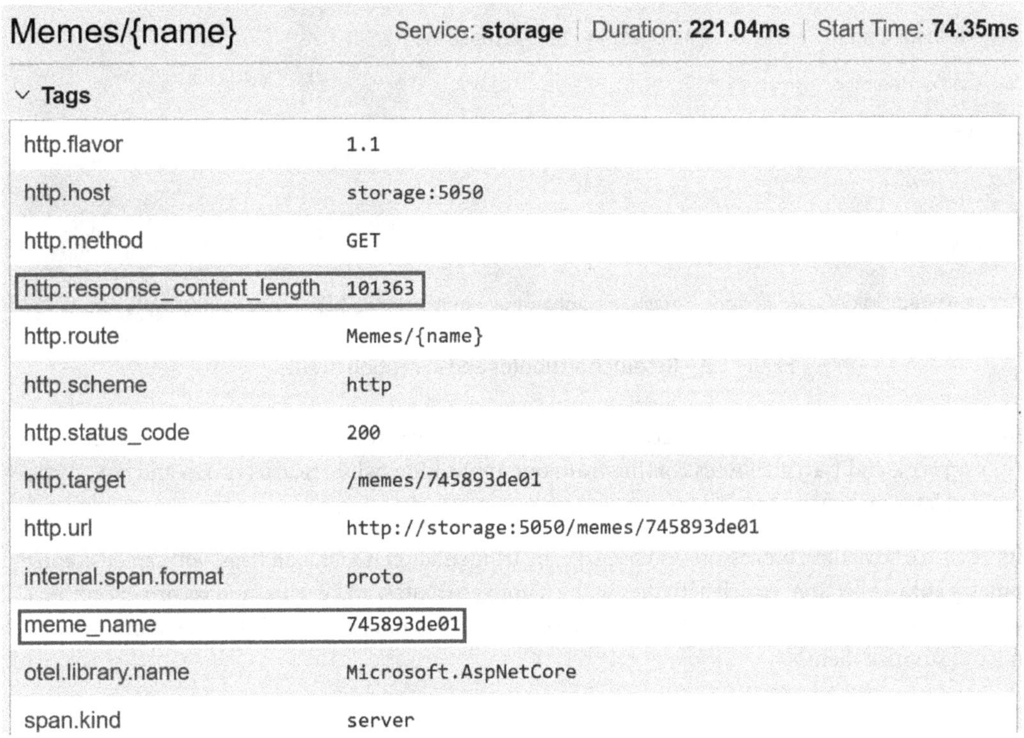

Memes/{name}	Service: **storage** Duration: **221.04ms** Start Time: **74.35ms**
∨ **Tags**	
http.flavor	1.1
http.host	storage:5050
http.method	GET
http.response_content_length	101363
http.route	Memes/{name}
http.scheme	http
http.status_code	200
http.target	/memes/745893de01
http.url	http://storage:5050/memes/745893de01
internal.span.format	proto
meme_name	745893de01
otel.library.name	Microsoft.AspNetCore
span.kind	server

Figure 5.2 – Auto-collected ASP.NET Core activity with custom attributes

We also get new resource attributes on every exported span, as well as exception events on incoming and outgoing HTTP spans, as shown in *Figure 5.3*:

Figure 5.3 – Resource attributes and exception events

Here, we can see `service.version`, `service.namespace`, `service.instance.id`, `env`, `region`, and `tag` attributes coming from our application, while `host.name` and `os.type` are added later on by the OpenTelemetry Collector resource detector.

With resource attributes, processors and baggage, instrumentation hooks, and flags, you can customize telemetry auto-collection, enrich activities with custom attributes, add events, and record exceptions. You can also change or remove attributes and suppress or filter out activities. But what about metrics, can we customize them?

Metrics

Auto-collected metrics are not as customizable as traces. With OpenTelemetry SDK, we can only enrich them using static resource attributes, but OpenTelemetry Collector provides processors that can add, remove, or rename attribute names and values, aggregate across attributes, change data types, or massage metrics in other ways.

Still, you can filter out specific instruments or their attributes using the `MeterProviderBuilder`. `AddView` method. For example, you can drop an instrument with a specific name using the following code:

Program.cs

```
WithMetrics(builder => builder.AddView(
    "process.runtime.dotnet.jit.il_compiled.size",
    MetricStreamConfiguration.Drop));
```

https://github.com/PacktPublishing/Modern-Distributed-Trac-ing-in-.NET/blob/main/chapter5/memes/frontend/Program.cs

Prometheus replaces the dot with an underscore and the corresponding instrument appears as `process_runtime_dotnet_jit_il_compiled_size` there. Check the corresponding OpenTelemetry instrumentation documentation to find the original name of the instrument. For example, .NET runtime instrumentation documentation can be found here: `github.com/open-telemetry/opentelemetry-dotnet-contrib/blob/main/src/OpenTelemetry.Instrumentation.Runtime/README.md`.

You can also specify attributes you want on the instrument – OpenTelemetry will use only those specified and will drop other attributes. It can be done to save costs on unused attributes or to remove high-cardinality attributes added by mistake.

For example, this code removes `http.scheme` and `http.flavor` from the ASP.NET Core request duration metric:

Program.cs

```
AddView("http.server.duration",
    new MetricStreamConfiguration(){
    TagKeys = new [" {"http.host", "http.method",
        "http.scheme", "http.target", "http.status_code" }

})
```

https://github.com/PacktPublishing/Modern-Distributed-Trac-ing-in-.NET/blob/main/chapter5/memes/frontend/Program.cs

We just saw how to enrich traces and metrics with application-specific context, and update or remove attributes using different mechanisms available in OpenTelemetry. Let's continue exploring OpenTelemetry configuration and learn how to configure context propagation.

Customizing context propagation

When instrumenting new systems, using W3C trace context propagation is the default and the easiest option – it does not need any explicit configuration since .NET and OpenTelemetry use it by default. However, existing systems may employ legacy context propagation conventions.

To support them, we can configure a custom global propagator on OpenTelemetry using `Sdk.SetDefaultTextMapPropagator`. For example, if one of your old client applications still uses some variation of custom correlation ID, you can still read it from request headers and convert it to a `trace-id`-compatible format (or move it to baggage).

You can use a composite propagator and support multiple standards at once as shown in this example:

XCorrelationIdPropagator.cs

```
Sdk.SetDefaultTextMapPropagator(
  new CompositeTextMapPropagator(new TextMapPropagator[] {
    new B3Propagator(true),
    new XCorrelationIdPropagator(),
    new BaggagePropagator()}));

DistributedContextPropagator.Current =
  DistributedContextPropagator.CreateNoOutputPropagator();
```

https://github.com/PacktPublishing/Modern-Distributed-Trac-ing-in-.NET/blob/main/chapter5/memes/frontend/XCorrelationIdProp-agator.cs

---

Here, we configured the B3 propagator from the `OpenTelemetry.Extensions.Propagators` package, a custom one for the `x-correlation-id` support, and also one for baggage. Note that we also disabled native ASP.NET Core and HTTP client propagation by setting `DistributedContextPropagator.Current` to the no-output propagator. If we don't do it, they will keep extracting and injecting `Trace-Context` headers.

When using composite context propagators, make sure to resolve collisions and define priority in case you get multiple conflicting combinations of trace context in the same request – we'll talk more about it in *Chapter 16, Instrumenting Brownfield Applications*.

Processing a pipeline with the OpenTelemetry Collector

As we've seen before, the OpenTelemetry Collector is another component that's capable of controlling, enriching, filtering, converting, routing, aggregating, sampling, and processing telemetry in any other possible way. *Figure 5.4* shows the main Collector components:

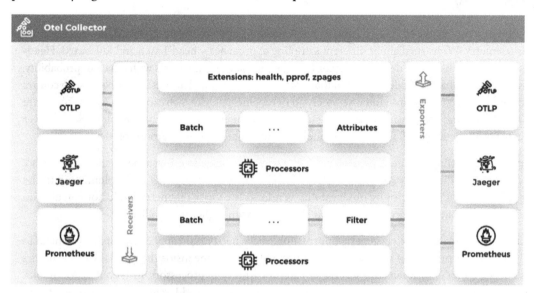

Figure 5.4 – OpenTelemetry Collector pipeline

Receivers get telemetry from different sources, and processors massage data and pass it over to exporters.

Since the collector is a separate process, potentially running on a different machine, it does not have a dynamic context, such as a specific HTTP request header, which you might want to stamp on spans. Such context can only be added inside your application.

But the collector could have more context about the environment the application runs in – for example, it can enrich telemetry with Kubernetes or cloud provider context. It can also receive telemetry in any format and convert it into OpenTelemetry signals.

The Collector supports a variety of receivers including Docker stats, statsd, or containerd for metrics, Kubernetes events, syslog, or journald for logs, and many others. Collector processors can enrich, filter, or change attribute names and values on any signal. One of the common use cases is sensitive data redaction.

You can find available Collector components in the OpenTelemetry Registry, `https://opentelemetry.io/registry`; just make sure to check the stability level for each component you consider using.

In many cases, it's not a question of whether to configure and customize your telemetry pipeline in-process or with Collector – it's both. Collector can help you migrate from one observability stack to another, provide safety belts on telemetry collection, and offload some work from your application.

Summary

In previous chapters, we explored telemetry auto-collection, and now we have learned how to customize this telemetry. We learned about different sampling approaches – head-based and tail-based. Head-based sampling makes the decision to record a trace (or span) when it starts with a certain probability. Children can follow parent decisions and then traces are always complete, but it's not possible to control the volume of traces on individual services. To overcome this, downstream services can configure different rates and use consistent sampling to maximize the number of complete traces. Some traces become partial but are still useful for monitoring individual services or groups of them.

Probability sampling captures a percentage of all traces and is great to mitigate performance overhead. If you need predictable costs, you should consider rate-based sampling. It's implemented in the OpenTelemetry Collector or by observability vendors. The OpenTelemetry Collector can also do tail-based sampling and record failures and long traces with higher probability.

Auto-instrumentations collect generic information, so we need to add application-specific context to gain a much deeper level of observability. Enrichment can be done inside the process with resources – static attributes describing your service instance, or with span processors, instrumentation hooks, or configuration options that can capture dynamic context. We can add, remove, or change attributes, and we can use baggage to propagate application-specific context within our system.

We can sometimes suppress activities with instrumentation hooks or flags or filter them out using processors. Filtering activities that have already started should be done with caution as it will break the correlation between dropped activity's ancestors and successors. It should only be done for activities that don't have children.

Metrics allow some customization too – we can enrich them with resource attributes, drop specific instruments, or limit attributes populated by instrumentations.

We also looked at context propagation customization, which can provide interoperability with custom and legacy correlation solutions you might have.

Finally, we talked about Collector features that can be used in addition to in-process configuration – environment resource detection or filtering and massaging telemetry. It can also provide rate-limiting sampling and protect your telemetry pipeline from overloading.

In this chapter, you learned to choose a sampling strategy and implement it, enrich traces with application context, and customize context propagation. It concludes our auto-instrumentation journey; from now on, we're going to explore instrumentation internals and learn how to write our own. In the next chapter, we'll focus on creating activities.

Questions

1. How would you build a general-purpose sampling solution that also captures failures and long-running traces in a distributed application?

2. How would you record retries on the HTTP client spans?

3. Configure rate-limiting sampling on the OpenTelemetry Collector.

6

Tracing Your Code

In the previous chapters, we talked about instrumentation libraries and learned how to use auto-collected telemetry to monitor and debug distributed systems. Auto-instrumentations, when available, provide necessary and reasonable coverage for network calls, but you may want to trace additional logical operations, I/Os, sockets, or other calls that don't have shared instrumentation available.

This chapter provides an in-depth guide to manual tracing using the `System.Diagnostics` primitives or the OpenTelemetry API and explains the mechanism behind auto-instrumentation. We will cover the `Activity` properties and how to populate them and show you how to record events. We'll then learn how to use links to represent complex relationships between spans. Finally, we'll cover the testing aspects of instrumentation.

You'll learn how to do the following:

- Create activities using the .NET API or with the OpenTelemetry API shim
- Use ambient context propagation with `Activity.Current` and understand its limitations
- Use `ActivityEvent` and understand when to use logs instead
- Represent complex relationships between traces with links
- Validate your instrumentations

By the end of this chapter, you should be able to cover your application needs with manual tracing.

Technical requirements

The code for this chapter can be found in this book's GitHub repository at `https://github.com/PacktPublishing/Modern-Distributed-Tracing-in-.NET/tree/main/chapter6`.

We'll need the following tools for it:

- NET SDK 7.0 or newer

- Visual Studio or Visual Studio Code with C# development setup is recommended, but any text editor will work

- Docker

If you want to explore traces from sample applications in Jaeger, you can run it with the following command:

```
$ docker run -d --name jaeger -p 6831:6831/udp -p
  16686:16686 jaegertracing/all-in-one:latest
```

Tracing with System.Diagnostics or the OpenTelemetry API shim

As we saw in the previous chapters, distributed tracing in .NET relies on primitives in the `System.Diagnostics` namespace: `Activity` and `ActivitySource`. They're used by HTTP Client and ASP.NET Core instrumentations.

The terminology used by .NET and OpenTelemetry is different: `Activity` represents the OpenTelemetry span, `ActivitySource` maps to the tracer, and tags map to attributes.

The `OpenTelemetry.Api` NuGet package also provides `TelemetrySpan`, `Tracer`, and several more auxiliary classes in the `OpenTelemetry.Trace` namespace. These APIs are a **shim** over .NET tracing APIs – a thin wrapper that does not provide any additional functionality on top of .NET tracing APIs.

You are probably wondering which one you should use. The rule of thumb is to use .NET APIs unless you want to stick to OpenTelemetry terminology. Shim is just an additional layer, which brings a small performance overhead.

Tracing with System.Diagnostics

Let's say we want to instrument an operation – for example, a method that processes a work item. It might break down into smaller, potentially auto-instrumented operations such as HTTP requests. In this case, spans describing individual requests won't show us the overall duration and result of processing, so we need to create a new logical `Activity` to describe it.

Activities should be created from `ActivitySource`, which notifies OpenTelemetry and other potential listeners about them. Using this approach, we can instrument our work processing operation with the following code:

Worker.cs

```
private static readonly ActivitySource Source =
  new ("Worker");
...
using var activity = Source.StartActivity("DoWork");
activity?.SetTag("work_item.id", workItemId);
try
{
  await DoWorkImpl(workItemId);
}
catch (Exception ex)
{
  activity?.SetStatus(ActivityStatusCode.Error,
    ex.Message);
  throw;
}
```

https://github.com/PacktPublishing/Modern-Distributed-Tracing-in-.NET/blob/main/chapter6/tracing-with-net/Worker.cs

In this example, the actual processing happens in the `DoWorkImpl` method. We created a new activity before calling into this method and implicitly ended it. If the method throws an exception, we set the status of the activity to `Error` while providing an exception message in the description. The activity is disposed of (and ended) when control leaves the scope, but we could also call the `Activity.Stop` method to stop it explicitly.

We created `ActivitySource` as a static singleton here since we're assuming we need it for the application's lifetime. If you decide to make it an instance variable and tie its lifetime to some long-living client or service in your application, make sure to dispose of it.

The only thing we configured in this example is the `Activity` name – we passed `DoWork` to the `ActivitySource.StartActivity` method.

> **Note**
>
> The activity returned by `ActivitySource.StartActivity` is nullable. It can be null if there are no listeners for this source, or if the listener sampled out this activity in a specific way by returning `ActivitySamplingResult.None`.

We'll learn more about APIs in a moment, but first, let's learn how to export generated activities.

Exporting activities with OpenTelemetry

So far, we have used the `OpenTelemetry.Extensions.Hosting` NuGet package to configure OpenTelemetry in ASP.NET Core applications. Here, we're going to use plain OpenTelemetry SDK configuration, which still looks quite similar to what we saw in *Chapter 5, Configuration and Control Plane*:

Program.cs

```
using var provider = Sdk.CreateTracerProviderBuilder()
  .ConfigureResource(b => b.AddService("activity-sample"))
  .AddSource("Worker")
  .AddJaegerExporter()
  .AddConsoleExporter()
  .Build()!;
```

https://github.com/PacktPublishing/Modern-Distributed-Tracing-in-.NET/blob/main/chapter6/tracing-with-net/Program.cs

In this example, we're building the `TracerProvider` instance by setting the service name to `activity-sample` and enabling `ActivitySource` with `Worker` – the one we used to create the activity in the previous example. We are also using the console exporter along with Jaeger – spans will be exported to both.

OpenTelemetry needs explicit configuration to listen to `ActivitySource`, but you can enable a group of them with wildcards, as we saw in *Chapter 3, The .NET Observability Ecosystem*.

Let's run these examples using the following command:

```
tracing-with-net$ dotnet run open-telemetry –scenario basic
```

We should see `Activity` exported to the console:

```
Activity.TraceId:           9c45e1b454e28bf1edbba296c3315c51
Activity.SpanId:            bcd47a4fc7d92063
Activity.TraceFlags:        Recorded
Activity.ActivitySourceName: Worker
Activity.DisplayName:       DoWork
Activity.Kind:              Internal
Activity.StartTime:         2022-12-07T23:10:49.1998081Z
Activity.Duration:          00:00:00.1163745
Resource associated with Activity:
    service.name: activity-sample
```

When we configure OpenTelemetry to listen to the `Worker` source, it leverages the `System.Diagnostics.ActivityListener` primitive. If you use OpenTelemetry, you probably won't need to use the listener directly, but you may still find it useful for testing purposes or when debugging instrumentation issues. Let's see how it works.

Listening to activities with ActivityListener

`ActivityListener` allows us to subscribe to any `ActivitySource` instance using its name and get notifications when activities created by one of the enabled sources start or end. The following example shows how to write a listener:

Program.cs

```
ActivitySource.AddActivityListener(new ActivityListener()
{
  ActivityStopped = PrintActivity
  ShouldListenTo = source => source.Name == "Worker",
  Sample = (ref ActivityCreationOptions<ActivityContext> _)
    => ActivitySamplingResult.AllDataAndRecorded
});
```

https://github.com/PacktPublishing/Modern-Distributed-Tracing-in-.NET/blob/main/chapter6/tracing-with-net/Program.cs

---

Here, we're subscribing to `ActivitySource` with `Worker` and specifying that we sample in all activities. When `Activity` ends, we call our `PrintActivity` method. We could also provide the `ActivityStarted` callback when needed.

So, let's go ahead and run this sample with this command:

```
tracing-with-net$ dotnet run activity-listener
```

You should see something like this:

```
DoWork: Id = 00-7720a4aca8472f92c36079b0bee3afd9-
0d0c62b5cfb15876-01, Duration=110.3466, Status = Unset
```

Now that you know how OpenTelemetry and `ActivitySource` work together, it's time to explore other tracing APIs.

Starting activities

The `ActivitySource` class defines several `CreateActivity` and `StartActivity` method overloads.

Calling into `StartActivity` is equivalent to calling `CreateActivity` and then starting it later with the `Activity.Start` method:

```
Source.CreateActivity("foo", ActivityKind.Client)?.Start()
```

The `Start` methods generate a new span ID, capture the start time, and populate the ambient context via the `Activity.Current` property. `Activity` can't be used until it's started. So, in most cases, the `StartActivity` method is the easiest choice and `CreateActivity` might only be useful if you want to construct an activity instance but start it later.

> **Note**
> Sampling callback happens during activity creation, so you must pass all properties that affect the sampling decision to the `StartActivity` or `CreateActivity` method.

Here are the start time properties:

- **Activity kind enumeration**: This is the same as the OpenTelemetry span kind. The default value is `Internal`, indicating local or logical operation. `Client` and `Server` spans describe the client and server side of synchronous remote calls such as HTTP requests. Similarly, the `Producer` and `Consumer` spans describe corresponding sides of asynchronous operations, such as asynchronous messaging.

  Observability backends rely on span kinds for visualizations such as service maps and semi-automated performance analysis.

- **Parent**: When instrumenting server or consumer calls, we extract parent trace context from headers into the `ActivityContext` struct. This is usually based on the W3C Trace Context standard for HTTP and might be different for other protocols. `ActivityContext` contains the trace ID, span ID, trace flags, and trace state.

  Another option is to pass the `traceparent` value in W3C Trace Context format to the `StartActivity` method as a string. You can set `tracestate` later after the activity starts, but then, of course, you won't be able to use it to make sampling decisions.

  If the parent context is not provided, `Activity.Current` is used.

- **Tags**: These are the attributes that are available when `Activity` starts and should affect sampling decisions. If you don't use attributes to make head-based sampling decisions, it's best not to populate them and minimize the performance overhead for sampled-out activities.

- **Links**: Links can correlate different traces and represent relationships between spans other than parent-child ones. We'll learn more about them later in this chapter.

- **Start time**: In some cases, we need to trace operations that happened in the past – for example, if we receive an event stating that something has ended and we want to convert it into an activity. Here, we can just fake the start time for this activity. We can also update it after `Activity` starts.

`Activity` also supports custom trace context formats – for example, legacy hierarchical ones.

After `Activity` starts, we can always add more attributes, change the start and end times, update the sampling decision, set `tracestate`, and record events.

Before adding new events or attribute, make sure to check the `IsAllDataRequested` flag, which specifies whether the activity has been sampled. We can use it to minimize the performance impact of the instrumentation by guarding any expensive operations:

StartSamples.cs

```
if (activity?.IsAllDataRequested == true)
    activity?.SetTag("foo", GetValue());
```

https://github.com/PacktPublishing/Modern-Distributed-Trac-ing-in-.NET/blob/main/chapter6/tracing-with-net/StartSamples.cs

The `Activity` and `ActivitySource` APIs are the foundation for any instrumentation. We'll cover additional APIs allowing to populate events and links later in this chapter and see more examples throughout the rest of this book. For now, let's take a quick look at how we can use the OpenTelemetry API shim.

Tracing with the OpenTelemetry API shim

The OpenTelemetry API shim does not provide any additional features on top of .NET APIs; it only aligns terminology with OpenTelemetry. If you use OpenTelemetry in other languages, then it might be more appealing to you. If you decide to go down this path, keep in mind that the behavior of `Tracer` and `Span` matches `ActivitySource` and `Activity`. For example, this means that you still have to enable each tracer when configuring OpenTelemetry.

Let's repeat our processing instrumentation using the `Tracer` and `TelemetrySpan` classes:

Worker.cs

```
private static readonly Tracer Tracer = TracerProvider
    .Default.GetTracer("Worker");
...
using var workSpan = Tracer.StartActiveSpan("DoWork"));
  workSpan.SetAttribute("work_item.id", workItemId);
  try
  {
    await DoWorkImpl(workItemId);
  }
  catch (Exception ex)
```

```
  {
    workSpan.SetStatus(Status.Error.WithDescription(
      ex.Message));
    throw;
  }
}
```

```
https://github.com/PacktPublishing/Modern-Distributed-Trac-
ing-in-.NET/blob/main/chapter6/tracing-with-shim/Worker.cs
```

The overall flow is the same: we create an instance of `Tracer` instead of `ActivitySource`, then use it to create a span. Adding attributes and setting the status is done similarly to the `ActivitySource` example.

If we were to look under the hood of `Tracer` and `TelemetrySpan`, we'd see that they fully rely on `ActivitySource` and `Activity`. As a result, enabling this instrumentation and enriching and customizing it is the same as enabling an `ActivitySource`-based one – it's done by using the `AddSource` method on `TracerProviderBuilder` (the source name matches the tracer name).

Even though the APIs look similar, there are a few important differences:

- Spans are not nullable. You always get an instance of a span, even if there is no listener to the underlying `ActivitySource` (but then it's an optimized, inoperative instance).

- All operations on spans are internally guarded with the `TelemetrySpan.IsRecording` flag, which is equivalent to the `activity?.IsAllDataRequested == true` check. However, it could still be useful to guard expensive operations to calculate attribute values and other span properties behind the `IsRecording` flag.

- Spans are not active (that is, current) by default. While you can't start an activity without making it current, that's not the case for `TelemetrySpan`. You probably noticed that we used the `Tracer.StartActiveSpan` method for the DoWork span, which populates `Activity.Current`.

  If we were to use the `Tracer.StartSpan` method, we'd get a started activity, but `Activity.Current` would not point to it. To make it current, we could call into the `Tracer.WithSpan` method.

If we run the previous OpenTelemetry API example with the `tracing-with-otel-api$ dotnet run` command, we'll see the same trace as before with plain .NET tracing APIs.

Now, let's see how we can create hierarchies of activities and enrich them using ambient context.

Using ambient context

In complex applications, we usually have multiple layers of spans in each trace. These spans are emitted by different libraries that are not aware of each other. Still, they are correlated because of the ambient context propagated in the `Activity.Current` property.

Let's create two layers of activities – we'll make processing more resilient by retrying failed operations and instrumenting tries and the logical `DoWork` operation:

Worker.cs

```
public static async Task DoWork(int workItemId) {
  using var workActivity = Source.StartActivity();
  workActivity?.AddTag("work_item.id",  workItemId);

  await DoWithRetry(async tryCount => {
    using var tryActivity = Source.StartActivity("Try");
    try
    {
      await DoWorkImpl(work.Id, tryCount);
      tryActivity?.SetTag("try_count", tryCount);
    }
    catch (Exception ex)
    {
      tryActivity?.RecordException(ex);
      tryActivity?.SetStatus(ActivityStatusCode.Error);
      throw;
    }
  }
}
```

https://github.com/PacktPublishing/Modern-Distributed-Trac-ing-in-.NET/blob/main/chapter6/tracing-with-net/Worker.cs

In this example, we have `workActivity`, which describes the logical operation, and `tryActivity`, which describes a try. Let's run it with the following command:

```
tracing-with-net$ dotnet run open-telemetry –scenario with-retries
```

Check out the trace in Jaeger at `http://localhost:16686`. You should see something similar to the trace shown in *Figure 6.1*:

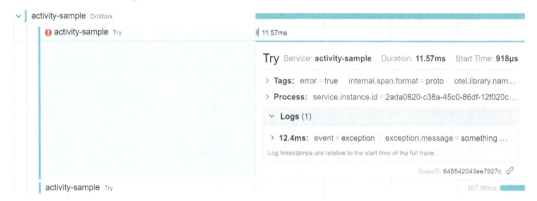

Figure 6.1 – Work item processing

Here, we can see that there were two attempts to process a work item – the first one failed with an exception and then the operation succeeded after the second try. By looking at this trace, it's clear why the `DoWork` operation took this much time – it was spent between tries.

Note that we didn't do anything special to correlate `workActivity` and `tryActivity`. This happened because `workActivity` was current when `tryActivity` started – since we didn't provide any parent, it defaulted to the `Activity.Current` instance.

To troubleshoot instrumentation issues, we can always check the parent of `Activity` on the started activity by looking at its properties. `Activity.Parent` represents an implicit parent. When we start an activity, we can also provide `traceparent` as a string or a parent `ActivityContext` explicitly – in these cases, the `Parent` property will be null. You can find some examples of this at `https://github.com/PacktPublishing/Modern-Distributed-Tracing-in-.NET/blob/main/chapter6/tracing-with-net/StartSamples.cs` and we'll see more examples of context propagation in *Chapter 10*, *Tracing Network Calls*. `Activity.ParentId` is the equivalent of the `traceparent` header and `Activity.ParentSpanId` represents the `span-id` portion of it. These properties are populated if `Activity` has any parent at all.

Getting back to our example, what happens if all the tries fail? Should we set an error on the `DoWork` activity? Well, we can do this inside the `DoWithRetry` method with `Activity.Current?.SetStatus(ActivityStatusCode.Error)`. We can use the current activity here because we control when and how the `DoWithRetry` method is called.

As a rule of thumb, avoid adding attributes, events, or setting a status on `Activity.Current` unless you know for sure it's the right one. With suppression, filtering, or some new activities created in between, `Current` can point to some other activity. So, make sure to pass instances of activities within your instrumentations explicitly.

If you want to enrich an auto-collected `Activity`, accessing the `Current` property in one of the enrichment callbacks provided when enabling instrumentation should be safe.

Some instrumentations may also provide access to the created `Activity`. For example, ASP.NET Core does so with the `IHttpActivityFeature` interface. It's also possible to walk up the activity tree using the `Parent` property to find the one you want to enrich.

`Activity.Current` works on top of `AsyncLocal`, so the .NET runtime propagates it through asynchronous calls. This does not work with background processing or manual manipulations with threads, but you can always pass activities explicitly and set the `Activity.Current` value manually as needed.

Now, we know how to create hierarchies of activities and use attributes to describe our scenarios. But sometimes, we need something more lightweight, such as events – let's take a closer look at them.

Recording events

Spans describe operations that have a duration and a result, but sometimes, creating a span could be too verbose and expensive – for example, for busy socket-level communication. Common use cases for events include recording exceptions or individual messages in gRPC streaming calls.

To represent something that happened at a certain point in time, we should use events or logs. In OpenTelemetry, the difference between logs and events is semantical – it's the same data structure with the same over-the-wire format but different attributes. For example, logs have mandatory severity, which does not apply to events. Events, on the other hand, have mandatory names.

They are also different in terms of their API and implementation (at least with .NET 7.0 and prior versions). In this section, we will explore Activity's events API; we will look at logs in *Chapter 8, Writing Structured and Correlated Logs*.

When to use events

To create an activity event, we need an instance of an activity, which is not the case, for example, at startup time.

Activity events depend on sampling – we can add them to a sampled-out `Activity`, but in general, they'll be dropped along with it.

An event's lifetime is tightly coupled with the `Activity` instance, so it'll stay in memory until it's garbage-collected. There is no limit regarding how many events you can have on the .NET side, but OpenTelemetry exporters limit the number of exported events. It is set to 128 by default and can be controlled with the `OTEL_SPAN_ATTRIBUTE_COUNT_LIMIT` environment variable.

> **Note**
>
> Activity events should be used to express operations that don't deserve a span, don't have a duration or are too short, and have a predictable outcome. Events must happen in the scope of some `Activity` and should only be exported if `Activity` is sampled in. There should also be a reasonable number of them under a single `Activity` instance. Given these limitations, logs are usually a better choice, as long as your observability backend supports them.

Now that we know these limitations, we're finally ready to play with events.

The ActivityEvent API

An event is represented with the `ActivityEvent` class. To create one, we must provide an event name, and can optionally specify a timestamp (that defaults to the time the event was constructed), as well as a collection of attributes.

The event's name is a low-cardinality string that implies the event's structure: events with the same name are expected to describe occurrences of the same thing and should use the same set of attributes.

Let's enrich HTTP client instrumentation with events. Imagine that we have read a long stream over HTTP and want to control content buffering.

To achieve this, we can pass the `HttpCompletionOption.ResponseHeadersRead` flag to the `HttpClient.SendAsync` method. The HTTP client will then return a response before reading the response body. It's useful to know the point in time when we got the response so that we know how long it took to read a response.

The following example demonstrates this:

Worker.cs

```
public static async Task DoWork(int workItemId) {
  using var work = Source.StartActivity();
  try
  {
    work?.AddTag("work_item.id", workItemId);
    var res = await Client.GetAsync(
      "https://www.bing.com/search?q=tracing",
      HttpCompletionOption.ResponseHeadersRead);
    res.EnsureSuccessStatusCode();

    work?.AddEvent(
      new ActivityEvent("received_response_headers"));
```

```
    ...
  }
  catch (Exception ex)
  {
    work?.SetStatus(ActivityStatusCode.Error,
      ex.Message);
  }
}
```

https://github.com/PacktPublishing/Modern-Distributed-Trac-
ing-in-.NET/blob/main/chapter6/events/Worker.cs

In this example, we start `Activity` to track overall logical request processing through the HTTP client pipeline and then record the `response_headers` event. This event does not have any attributes – its sole purpose is to record the timestamp when we got a response from the server.

Let's add more events! Assuming we use throttling or circuit-breaking in our HTTP pipeline, we won't have any physical HTTP requests and no spans reported by auto-instrumentation. Events can provide observability into it.

We'll implement client-side throttling using the `RateLimiter` class, which is available in .NET 7 and included in the `System.Threading.RateLimiting` NuGet package. We'll do so in the `DelegatingHandler` class, as shown in this example:

RateLimitingHandler.cs

```
private readonly TokenBucketRateLimiter _rateLimiter =
  new (Options);
protected override async Task<HttpResponseMessage>
  SendAsync(HttpRequestMessage req, CancellationToken ct)
{
  using var lease = _rateLimiter.AttemptAcquire();
  if (lease.IsAcquired)
    return await base.SendAsync(req, ct);

  return Throttle(lease);
}
```

https://github.com/PacktPublishing/Modern-Distributed-Trac-
ing-in-.NET/blob/main/chapter6/events/RateLimitingHandler.cs

Here, we are trying to acquire a lease from the rate limiter. If it's successfully acquired, we call into the `base.SendAsync` method, letting this request process further. Otherwise, we must throttle the request, as demonstrated in the following code snippet:

RateLimitingHandler.cs

```
private HttpResponseMessage Throttle(RateLimitLease lease)
{
  var res = new HttpResponseMessage(
    HttpStatusCode.TooManyRequests);
  if (lease.TryGetMetadata(MetadataName.RetryAfter,
    out var retryAfter))
  {
    var work = Activity.Current;
    if (work?.IsAllDataRequested == true)
    {
      var tags = new ActivityTagsCollection();
      tags.Add("exception.type", "rate_is_limited");
      tags.Add("retry_after_ms",
        retryAfter.TotalMilliseconds);
      work?.AddEvent(new ActivityEvent("exception",
        tags: tags));
    }
    res.Headers.Add("Retry-After",
      ((int)retryAfter.TotalSeconds).ToString());
  }
  return res;
}
```

https://github.com/PacktPublishing/Modern-Distributed-Tracing-in-.NET/blob/main/chapter6/events/RateLimitingHandler.cs

In the `Throttle` method, we emit the `exception` event and provide a message alongside the `retry_after` attribute. We got a value of this attribute from the rate limiter; it provides a hint regarding when it will make sense to retry this request.

The example in the `events` folder demonstrates a full rate-limiting solution – it configures the rate limiter to allow one request every 5 seconds but sends two requests in parallel so that the first one comes through and the second one is throttled.

Go ahead and run the sample with `events$ dotnet run` and then switch to Jaeger to see two traces from the `events-sample` service.

One trace has two spans and represents a successful operation, as shown in *Figure 6.2*:

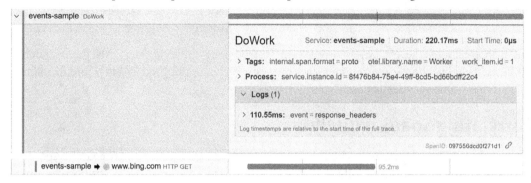

Figure 6.2 – A trace with logical and physical HTTP spans and the response_headers event

Here, we can see that the time to first byte was around 110 milliseconds. Then, we got the `response_headers` event; the rest of the logical `DoWork` operation was spent on reading stream contents.

The other trace has just one span and represents a failed operation; it's shown in *Figure 6.3*:

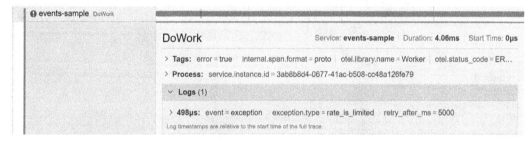

Figure 6.3 – A trace with a logical call and the rate_is_limited exception event

Here, we can see a logical `DoWork` span, which ended with an error. If we expand the attributes, we'll see a status description stating `Response status code does not indicate success: 429 (Too Many Requests)`. This could give us an idea of what happened, even if we didn't have an event. There is no physical HTTP span here and it could be confusing and unclear where the response came from.

With the `rate_is_limited` event, we can populate additional attributes such as `retry_after_ms`, but most importantly, we can easily understand the root cause of the problem and find the place in our code where the event is sent from.

Recording exceptions

In the previous example, we created an event that represents an error, which is a special event defined in OpenTelemetry. It has `exception` as its name and the `exception.type`, `exception.message`, and `exception.stacktrace` attributes. Either `type` or `message` is required.

If we had an exception object, we could have used a `RecordException` extension method declared in the `OpenTelemetry.Trace.ActivityExtensions` class. We could record exceptions using `activity?.RecordException(ex)` and then pass custom tag collection to add to the event.

This method calls into the `Activity.AddEvent` method under the hood, filling in all the exception attributes, including the stack trace. Since stack traces can be huge, it's a good idea to record them for unhandled exceptions and only once.

Correlating spans with links

So far, we have talked about parent-child relationships between spans. They cover request-response scenarios well and allow us to describe distributed call stacks as a tree, where each span has at most one parent and as many children as needed.

But what if our scenarios are more complicated? For example, how do we express receiving temperature data from multiple sensors and aggregating it on the backend, as shown in *Figure 6.4*?

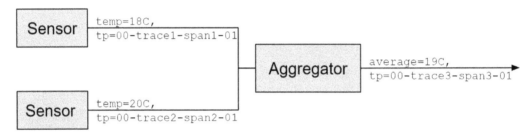

Figure 6.4 – Batch processing

In this example, sensors send data to the aggregator in the scope of different traces. The aggregator must start a third one – it shouldn't continue one of the sensor's traces.

We can use links to connect `trace3` to both `trace1` and `trace2`, allowing us to correlate all of them. Links don't specify exact relationships between spans, but in the scope of this example, we can think about them as multiple parents for a single span.

Links are mostly used in messaging scenarios where messages are sent and received in batches to optimize network usage, or could be also processed together.

Links have two properties: a linked trace context and a collection of attributes. Currently, they can only be provided to the `StartActivity` method and may be used to make sampling decisions. This is an OpenTelemetry specification limitation that might be removed in the future.

Using links

Let's see how we can use links to instrument batch processing scenarios with an in-memory queue. With background processing, we can't rely on `Activity.Current` to flow from the enqueue operation to processing. So, we'll pass `ActivityContext` along with the work item through the queue.

But first, we need to create an `Activity` for the enqueue operation so that we have some context to capture and pass around:

Producer.cs

```
public void Enqueue(int id)
{
  using var enqueue = Source
    .StartActivity(ActivityKind.Producer)?
    .SetTag("work_item.id", id);
  _queue.Enqueue(new WorkItem(id, enqueue?.Context));
}
```

https://github.com/PacktPublishing/Modern-Distributed-Tracing-in-.NET/blob/main/chapter6/links/Producer.cs

While it's important to instrument publish calls for remote queues, it's not essential in this example. We only did it here to have some valid `ActivityContext` captured. If we had any other activity, we could use its context instead.

Now, we're ready to instrument the work item processor:

BatchProcessor.cs

```
async Task ProcessBatch(List<WorkItem> items)
{
  using var activity = Source.StartActivity(
      ActivityKind.Consumer,
      links: items
        .Select(i => new ActivityLink(i.Context)));

  activity?.SetTag("work_items.id",
      items.Select(i => i.Id).ToArray());
  ...
}
```

https://github.com/PacktPublishing/Modern-Distributed-Tracing-in-.NET/blob/main/chapter6/links/BatchProcessor.cs

Here, we iterated over work items and created an `ActivityLink` for each of them by using the trace context passed alongside the `WorkItem` instance.

Then, we added the `work_item.id` attribute with an array containing all received IDs to the `BatchProcessing` activity. Ideally, we'd put attributes on the links themselves via the `ActivityLink` constructor, but I'm not aware of any observability backend that supports it now. As an alternative, we can also create an event for each work item and populate attributes on them.

Let's run the sample with `links$ dotnet run`. It will enqueue three work items and then process them all in one batch. In Jaeger, we should see four independent traces – one for each enqueue operation and one for batch processing. An example of the latter is shown in *Figure 6.5*:

Figure 6.5 – Processing span with links

We can see that it has three references (links in Jaeger terminology), which we can click on and land on the corresponding `Enqueue` operation, as shown in *Figure 6.6*:

Figure 6.6 – Enqueue span

In Jaeger, it's not possible to navigate from the `Enqueue` span to the `ProcessBatch` span. But some of the observability backends support navigation in both directions. For example, *Figure 6.7* shows the `Enqueue` operation linked to processing in Azure Monitor:

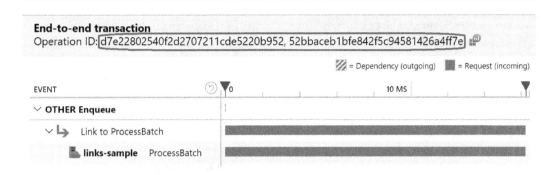

Figure 6.7 – Two linked traces visualized in Azure Monitor

Note that there are two different operations IDs (trace IDs) that have been correlated using a link. We'll see more examples of links in messaging scenarios in *Chapter 11, Instrumenting Messaging Scenarios*. For now, let's learn how to test our instrumentations.

Testing your instrumentation

The idea of testing logs might look wild – logs are not intended to stay in one place or retain a specific structure. It's not the case for traces.

Instrumentation directly affects your ability to evaluate production health and usage. Testing auto-instrumentation could be limited to basic happy case validation – we only need to check that it's enabled and emits some data in the right format. This would help us detect potential problems with dependency updates. Manual instrumentation needs more attention.

Let's see how we can test any instrumentation in ASP.NET Core applications. We're going to rely on the integration testing capabilities provided by the `Microsoft.AspNetCore.Mvc.Testing` NuGet package. You can find more details about it in the ASP.NET Core testing documentation available at `https://learn.microsoft.com/aspnet/core/test/integration-tests`. It allows us to modify the ASP.NET Core application's configuration for test purposes. In this section, we'll use it to change the OpenTelemetry pipeline and intercept activities.

Intercepting activities

There are a few different ways to intercept activities; let's list them:

- We can add a span processor, similar to how we enriched activities in *Chapter 5, Configuration and Control Plane*. Since processors run synchronously, we can validate the `Activity` attributes against the ambient context – for example, `Baggage.Current`. We can also check (when needed) that attributes are provided at start time in the `OnStart` callback.

- We can implement a test exporter. The downside of this approach is that we'll only see completed activities. Also, exporters run asynchronously and there will be no ambient context to validate against.

- We can write a custom `ActivityListener` implementation. This approach would not allow us to test the customization and configurations we've done with OpenTelemetry. We won't even be able to validate whether OpenTelemetry is configured to listen to specific `ActivitySource` instances or check whether sampling works as expected.

So, `ActivityListener` could be a great choice for unit testing, and the processor gives the most flexibility in terms of integration testing, which we're going to focus on here. Let's see how we can inject a processor into the OpenTelemetry pipeline in tests.

The `OpenTelemetry.Extensions.Hosting` NuGet package allows us to customize the pipeline with the `ConfigureOpenTelemetryTracerProvider` extension method. It's called after the OpenTelemetry pipeline is configured, right before the `TracerProvider` instance is built. If you are using vanilla OpenTelemetry, you will have to implement a callback for tests to alter the pipeline.

Here's an example of adding a test processor:

TestFactory.cs

```
public class TestFactory : WebApplicationFactory<Program>
{
  public readonly TestActivityProcessor Processor = new ();
  protected override void ConfigureWebHost(
    IWebHostBuilder b)
  {
    b.ConfigureServices(s => {
      s.ConfigureOpenTelemetryTracerProvider(
        (_, traceProviderBuilder) =>
          traceProviderBuilder.AddProcessor(Processor));
      ...
    });
  }
}
```

https://github.com/PacktPublishing/Modern-Distributed-Tracing-in-.NET/blob/main/chapter6/testing/app.tests/TestFactory.cs

The `TestFactory` class allows us to set up an ASP.NET Core application for testing, which we're doing in the `ConfigureWebHost` method. There, we call into the `ConfigureOpenTelemetryTracerProvider` method, where we change the OpenTelemetry pipeline and inject our test processor. Here's the minimalistic processor implementation:

TestActivityProcessor.cs

```
public class TestActivityProcessor :BaseProcessor<Activity>
{
  ConcurrentQueue<Activity> _processed = new ();
  public override void OnEnd(Activity activity) =>
    _processed.Enqueue(activity);

  ...
}
```

https://github.com/PacktPublishing/Modern-Distributed-Tracing-in-.NET/blob/main/chapter6/testing/app.tests/TestActivityProcessor.cs

We're almost ready to write some tests, but there is another challenge – how do we filter activities related to a specific test?

Filtering relevant activities

When we run tests in parallel, we effectively register multiple OpenTelemetry pipelines that listen to the same `ActivitySource` instances. With unit tests covering our instrumentations, we can control this better, but in the case of integration tests, `ActivitySource` and its listeners are de facto static and global – if we run tests in parallel, we'll see activities from all of them in the processor. We need to filter relevant activities that belong to our test, which we can do using distributed tracing.

We'll start a new activity for each test and propagate the context to the service under test. Then, we can filter processed activities based on their trace IDs. This approach is implemented in `TracingTests` (https://github.com/PacktPublishing/Modern-Distributed-Tracing-in-.NET/blob/main/chapter6/testing/app.tests/TracingTests.cs).

Validation

Once we can filter all the activities related to this test execution, we're ready to do some checks on them. It's useful to check all the properties you rely upon in your monitoring and debugging tasks. For example, the following code validates a few properties of ASP.NET Core's `Activity`:

TracingTests.cs

```
Assert.Equal("/document/foo",
  httpIn.GetTagItem("http.target"));
Assert.Equal(404, httpIn.GetTagItem("http.status_code"));
Assert.Equal(ActivityStatusCode.Unset, httpIn.Status);
Assert.Empty(httpIn.Events);
```

https://github.com/PacktPublishing/Modern-Distributed-Trac-ing-in-.NET/blob/main/chapter6/testing/app.tests/TracingTests.cs

Now, you're ready to write and test your instrumentations! You may also find it useful to use distributed tracing for your general integration testing needs – relying on it to validate intended test behavior and investigate flaky tests or unstable service behavior. You could also use traces as one of the inputs to validate service behavior.

Summary

In this chapter, we explored manual distributed tracing instrumentation using .NET diagnostics primitives. `Activity` and `ActivitySource` are the default ways to instrument your code – create, start, end, and enrich activities with attributes and events. You can achieve the same functionality with `Tracer` and `TelemetrySpan`, from the OpenTelemetry API package. They provide a thin wrapper over .NET diagnostics APIs while using OpenTelemetry terminology.

We also looked into the ambient context propagation with `Activity.Current` and how it makes multiple instrumentation layers work together. Then, we learned about events and their limitations and used links to correlate different traces.

Finally, we covered testing – since instrumentation can be critical for monitoring, we should validate it as any other feature. We learned how to reliably do this in ASP.NET Core applications.

With this, you should be able to write rich tracing instrumentations and troubleshoot and validate custom tracing code. To achieve better observability, we can combine multiple signals with minimal duplication, so in the next chapter, we're going to look at manual metrics instrumentation and see how it can work along with tracing.

Questions

1. Let's say you started `Activity` using `ActivitySource`. How do you configure OpenTelemetry to listen to it? How does it work?

2. When should you use `Activity` events? What are the alternatives?

3. What do we need links for? How can we use them?

7

Adding Custom Metrics

In the previous chapter, we looked into manual distributed tracing instrumentation, which should help us debug individual operations or analyze service usage with ad hoc queries. Here, we'll talk about metrics. First, we'll learn when to use them, understand cardinality requirements, and then see how traces and metrics complement each other. We'll explore the metrics API's evolution in .NET and then spend most of this chapter talking about OpenTelemetry metrics. We'll cover **instruments** such as counters, gauges, and histograms, and learn about each of them in depth.

We will cover the following topics in the chapter:

- The benefits, limitations, and evolution of metrics in .NET

- How and when to use different counters

- How to record and use gauges

- How to record value distribution with histograms

By the end of this chapter, you should be able to pick the right instrument for each scenario and implement and use it in your applications to analyze performance, health, and usage.

Technical requirements

The code for this chapter is available in this book's repository on GitHub at `https://github.com/PacktPublishing/Modern-Distributed-Tracing-in-.NET/tree/main/chapter7`.

To run the samples and perform analysis, we'll need the following tools:

- .NET SDK 7.0 or later

- Docker and `docker-compose`

Metrics in .NET – past and present

Even though we are focusing on distributed tracing in this book, learning about metrics is important to understand when and how to use them to improve observability.

Metrics allow us to report data that's been aggregated over a certain period and set of attributes (that is, dimensions or tags). A metric can be represented as a set of time series where each series measures the change of one indicator with a unique combination of attribute values over time. Examples include CPU utilization for a given service instance or HTTP request latency for a given route, HTTP method, response code, and instance.

The key difference between traces and metrics is aggregation – traces capture individual operations with detailed attributes. Traces answer questions such as *"What happened with this specific request?"* and *"Why did it happen?"* Metrics, on the other hand, tell us what happens in the system or specific parts of it, how common a failure is, how widespread the performance issue is, and so on.

Before diving into the use cases, benefits, and APIs of metrics, we first need to learn about the main limitation of metrics – low **cardinality**.

Cardinality

Cardinality represents number of combinations of unique attributes or number of time series. Adding a new attribute causes a combinatorial explosion of a time-series number, which leads to the combinatorial growth of a metric's volume.

> **Note**
>
> Metrics should have low cardinality, but "low" and "high" are relative – their definition depends on the budget, backend limits, and local memory consumption.

For example, a relatively big Prometheus instance can support millions of active time series. If we have 1,000 instances of a service running, and the service exposes four HTTP routes with three methods each and returns five different status codes, we're going to report 1,000 (instances) * 4 (routes) * 3 (methods) * 5 (status codes) = 60K time series of an HTTP server's request duration metric (at worst). If we try to include something such as the customer_id attribute and have 1,000 active customers, we'll start reporting 60M time series for the HTTP server request duration metric only.

We can still do this by scaling Prometheus horizontally, so reporting a few high-cardinality attributes is still feasible when justified.

> **Note**
>
> Metrics are aggregated in memory before they are exported, so metrics with high cardinality may affect application performance.

There are no limits in terms of attribute cardinality in .NET, but the OpenTelemetry SDK has configurable limits on the maximum number of metrics and the number of attribute combinations per metric.

When to use metrics

Resource consumption, low-level communication details, or the number of open connections are best represented by metrics. In other cases, we have a choice and can report telemetry as metrics, spans, or events. For example, if we want to measure number of incoming HTTP requests by route, we can query spans that are filtered by service, route, and timestamp. Should we also report metrics for it? Let's see.

Metrics are implemented and optimized under the assumption of low cardinality, which enables several important benefits:

- **Predictable costs and limited resource consumption**: The metrics' volume does not grow much as load increases – we only get a new set of time series when the service scales up adding new instances.
- **Low performance impact**: Reporting a single measurement can be done without allocating memory.
- **Unbiased usage and performance data**: Metrics are recorded regardless of sampling decisions. Metrics don't always report exact data, but we can control their precision by configuring collection intervals and histogram boundaries.
- **Fast and cheap(er) queries**: While observability backends store metrics in different ways and their pricing options vary, metrics are much more compact, which usually leads to faster ingestion and cheaper queries.

Metrics work best when we use them to monitor service health and usage regularly.

When you want to instrument some operation and are in doubt about which signal to use, the following strategy can help: if you need unbiased data or want to create an alert or a chart on a dashboard, use metrics. Otherwise, start with tracing and ad hoc queries. If you find yourself running a lot of similar queries over traces, then add a metric to optimize such queries.

Assuming your tracing backend does not support rich queries, you probably want to be more proactive in adding metrics. And if your backend is optimized for high-cardinality data and ad hoc analysis, you might not need metrics much.

Now that we have a rough idea of when we need metrics, let's dive into instrumentation.

Reporting metrics

There are a few different metrics (and counters) APIs in .NET – let's take a look at them and learn when to use them.

Performance counters

The System.Diagnostics.PerformanceCounter class and its friends implement Windows performance counters. They don't support dimensions. These limitations make performance counters an unlikely choice for a modern distributed system monitoring story.

Event counters

System.Diagnostics.Tracing.EventCounter is a cross-platform implementation of a counter, which represents a single time series – we saw it in action in *Chapter 2*, *Native Monitoring in .NET*, and *Chapter 4*, *Low-Level Performance Analysis with Diagnostic Tools*, where we collected counters coming from .NET with dotnet-counters and dotnet-monitor. OpenTelemetry can listen to them too, converting them into OpenTelemetry metrics and enriching them with resource attributes.

If you want to report a metric that does not need any dimensions except static context and want to be able to dynamically turn this metric on and off using diagnostics tools, event counters would be a good choice.

We're not going to dive into the EventCounter API, so please refer to the .NET documentation (https://learn.microsoft.com/dotnet/core/diagnostics/event-counters) to find out more.

OpenTelemetry metrics

The APIs we're going to focus on are available in the System.Diagnostics.Metrics namespace in the System.Diagnostics.DiagnosticSource NuGet package. These APIs follow OpenTelemetry's metrics specification and terminology, except the term "tags" is used instead of "attributes." There is no shim for metrics.

The metrics API supports recording multi-dimensional data using the following instruments:

- **Counter**: Represents a value that increments over time – for example, the number of open connections

- **UpDownCounter**: Represents an additive value that increments or decrements over time – for example, the number of active connections

- **Gauge**: Represents a current value – for example, the sequence number of the last message received from the messaging queue

- **Histogram**: Represents a distribution of value – for example, HTTP request latency

Instruments can be created with the Meter class. So, first, we need an instance of Meter, which we can create using a name and optional instrumentation version: Meter meter = new ("sample").

The name of `Meter` can match the application name, namespace, class, or anything else that makes sense in your case. It's used to enable metrics, as shown in the following example:

Program.cs

```
using var meterProvider = Sdk.CreateMeterProviderBuilder()
  .AddMeter("queue.*")
  .AddOtlpExporter()
  .Build()!;
```

`https://github.com/PacktPublishing/Modern-Distributed-Trac-ing-in-.NET/blob/main/chapter7/metrics/Program.cs`

Here, we enabled all metrics coming from any `Meter` whose name starts with `queue.` (we can use an exact match or wildcards).

`Meter` is disposable. In some cases, when you use the same `Meter` instance for the application's lifetime, you can make `Meter` instances static; otherwise, make sure to dispose of them to disable all nested instruments.

> **Note**
>
> We can listen to metrics directly, without OpenTelemetry, using the `System.Diagnostics.Metrics.MeterListener` class. It can subscribe to specific instruments and record their measurements. `MeterListener` is used by OpenTelemetry, so you might find it useful to debug instrumentation.

Now that we have a `Meter` instance and configured OpenTelemetry to export metrics, we can create instruments using factory methods on the `Meter` class; for example, `meter.CreateCounter<long>("connections.open")`.

We'll see how to create instruments later in this chapter, but for now, here's a list of common instrument properties:

- **A type parameter** represents a value type that must be a primitive number (`byte`, `short`, `int`, `long`, `float`, `double`, or `decimal`).
- **The instrument name** represents a unique exported metric name; it's a required property. OpenTelemetry limits the instrument name to 63 characters and has other limitations. We'll talk about it more in *Chapter 9, Best Practices*.
- **A unit** represents an optional value unit following Unified Code for Units and Measure (`https://unitsofmeasure.org/`).
- **The description** is an optional free-form piece of text that briefly describes the instrument.

We can create multiple instances of instruments with the same name in the process – the OpenTelemetry SDK aggregates data coming from them into one value. Instruments are de facto identified by their name, unit, and combination of resource attributes. So, measurements from multiple instrument instances that share the same identity are aggregated together. Let's explore instruments one by one and learn how to use them, starting with counters.

Using counters

Counter and **UpDownCounter** represent additive values – values that it makes sense to sum up. For example, the sum of incoming request counts with different HTTP methods makes sense, but the sum of CPU utilization across different cores does not.

On the instrumentation side, the only difference between **Counter** and **UpDownCounter** is that the former increases monotonically (or stays the same), while the latter can decrease. For example, the number of open and closed connections should be represented by **Counter**, while the number of active connections should be represented by **UpDownCounter**.

Both kinds of counters can be synchronous and asynchronous:

- **Synchronous** counters report a delta of value when change happens. For example, once we successfully initiate a new connection, we can increment counters for both open and active connections. Once we've finished terminating a connection, we decrement the number of active connections only.

- **Asynchronous** (that is, **observable**) counters are reported in a callback that's initiated periodically by the OpenTelemetry SDK. Such instruments return a value at call time. For example, when reporting the in-memory queue length, we can either use the UpDownCounter instrument and increment it when the item is enqueued and decrement when dequeued or create ObservableUpDownCounter and return the queue's length in a callback.

Let's instrument in-memory queue processing and learn about each instrument as we go.

The Counter class

A synchronous counter is implemented in the System.Diagnostics.Metrics.Counter class. We'll use it to keep track of the number of enqueued items:

Producer.cs

```
private static Meter Meter = new("queue.producer");
private static Counter<long> EnqueuedCounter =
  Meter.CreateCounter<long>("queue.enqueued.count",
```

```
    "{count}",
    "Number of enqueued work items");
```

https://github.com/PacktPublishing/Modern-Distributed-Trac-
ing-in-.NET/blob/main/chapter7/metrics/Producer.cs

Here, we created an instance of the `Meter` class called `queue.producer`. It's static here because we never need to disable corresponding instruments. Then, we created a static counter called `queue.enqueue.count` with a `long` type parameter, with the unit set to `{count}`.

We also need to increment it every time an item is enqueued. `Counter` exposes the `Add` method to record a positive delta; it has several overloads to pass zero or more attributes. In our sample, we have multiple queues and pass queue names:

Producer.cs

```
EnqueuedCounter.Add(1,
  new KeyValuePair<string, object?>("queue", _queueName))
```

https://github.com/PacktPublishing/Modern-Distributed-Trac-
ing-in-.NET/blob/main/chapter7/metrics/Producer.cs

Let's run the sample application with `metrics$ docker-compose up --build` and open the metrics endpoint on OpenTelemetry Collector at `http://localhost:8889/metrics`. We should see `queue_enqueued_count_total` among other metrics in the Prometheus exposition format:

```
# HELP queue_enqueued_count_total Number of enqueued work items
# TYPE queue_enqueued_count_total counter
queue_enqueued_count_total{job="metrics ",queue="add"} 323
queue_enqueued_count_total{job="metrics ",queue="remove"} 323
```

Here, we can see the description, along with the instrument's type, followed by a list of all attribute combinations and the latest reported counter values.

We can also visualize this counter in Prometheus (at `http://localhost:9090`). Usually, we are interested in rates or trends, and not the absolute value of a counter. For example, the rate at which items are enqueued would be a good indication of producer load and performance.

We can get this by using the `sum by (queue) (rate(queue_enqueued_count_total[1m]))` query – Prometheus calculates the rate per second (and averages it over 1 minute), then sums up values by grouping them by queue name across application instances. The corresponding chart is shown in *Figure 7.1*:

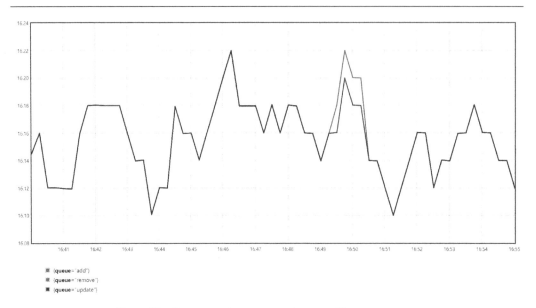

Figure 7.1 – Enqueue rate per second grouped by queue name

Here, we can see that we enqueue items at around 16 items per second to each queue. Instead of sum, we could use the min or max operators to see whether there are application instances that stand out.

Counters, along with other instruments, expose an Enabled flag, which indicates whether there are any listeners for this instrument. Meters are not enabled by default and the specific instrument could be disabled, so the Enabled flag should be used to guard any additional work necessary for metric reporting. It's really important for native instrumentations, where end users of such libraries may or may not have metrics enabled and the goal is to have zero performance impact when metrics are disabled.

Other properties that are exposed on the instruments include Name, Unit, Description, and Meter, which we used to create this instrument.

The UpDownCounter class

The System.Diagnostics.Metrics.UpDownCounter class is very similar to Counter in terms of the API. You can create one with the CreateUpDownCounter method on the Meter instance by providing an instrument name, along with an optional unit and description. The UpDownCounter class exposes an Add method, which takes a delta of the measured value and zero or more tags. It also exposes the Enabled flag and the properties the instrument was created with, such as its name, unit, and description.

On the consumption side, however, UpDownCounter is different. It's not monotonic and maps to the gauge type in Prometheus. We'll learn more about it in *The ObservableUpDownCounter class* section.

The ObservableCounter class

`System.Diagnostics.Metrics.ObservableCounter` implements an asynchronous version of `Counter`. There is no difference between synchronous and asynchronous counters on the consumption side. `ObservableCounter` just provides a more convenient way to record counters in a callback executed periodically.

For example, the number of completed (by the thread pool) tasks (`process.runtime.dotnet.thread_pool.completed_items.count`) available in the `OpenTelemetryInstrumentation.Runtime` NuGet package is implemented as `ObservableCounter`. On every call, it returns the `ThreadPool.CompletedWorkItems` property.

We can create an observable counter with the `CreateObservableCounter` method: `Meter.CreateObservableCounter<long>("my.counter", GetValue)`. Here, in addition to the name, we pass a lambda function – `GetValue` – which returns the current value of the counter.

It's executed when metrics are about to be exported. In our application, this happens every 5 seconds, while the default period for the OTLP exporter is 60 seconds. We configured it with the `OTEL_METRIC_EXPORT_INTERVAL` environment variable, but it's also possible to set it explicitly with the `PeriodicExportingMetricReaderOptions.ExportIntervalMilliseconds` property:

ExplicitConfiguration.cs

```
AddOtlpExporter((exporterOptions, readerOptions) =>
  readerOptions.PeriodicExportingMetricReaderOptions
    .ExportIntervalMilliseconds = 5000)
```

https://github.com/PacktPublishing/Modern-Distributed-Tracing-in-.NET/blob/main/chapter7/metrics/ExplicitConfiguration.cs

The `ExportIntervalMilliseconds` property controls how frequently counter values are collected, so it controls the precision and volume of individual time series.

This configuration does not affect pull-based exporters such as Prometheus, where it's controlled externally (for example, with the `scrape_interval` parameter on the Prometheus instance). In our sample application, we have the OTLP exporter, which is push-based and sends metrics to OpenTelemetry Collector. Collector then exposes metrics on the `http://localhost:8889/metrics` endpoint, where Prometheus scrapes them from.

With `ObservableCounter`, we can only record data using the callback provided at start time, and there are several overloads of the `CreateObservableCounter` method that allow us to report the metric's value, along with its attributes (via the `Measurement` struct) or as a list of `Measurement` instances.

There are several important things to know about the callback:

- Unlike the `Counter.Add` method, it reports an absolute value of the counter.

- It should finish in a reasonable amount of time. We can configure the timeout similarly to the export interval with the `OTEL_METRIC_EXPORT_TIMEOUT` environment variable or the `PeriodicExportingMetricReaderOptions.ExportTimeoutMilliseconds` property.

- The callback should not return multiple measurements for the same set of attributes. The OpenTelemetry SDK's behavior is not defined for this case.

> **Note**
>
> These requirements are from the OpenTelemetry specification. `MeterListener` does not enforce any of them.

To unsubscribe from observable counters, we must dispose of the corresponding `Meter` instance. So, if the counter relies on any instance data and belongs to an object with a limited lifetime, we must create `Meter` as an instance variable and dispose of it, along with the object it belongs to. Let's see an example of this with `ObservableUpDownCounter`.

The ObservableUpDownCounter class

`System.Diagnostics.Metrics.ObservableUpDownCounter` represents an asynchronous version of `UpDownCounter`. Its creation is similar to `ObservableCounter`, but its consumption side matches `UpDownCounter`.

We'll use it to report queue length – it should give us a good indication of processor throughput and whether it processes items fast enough.

The queue length is not monotonic – it can go up and down, so a regular counter would not work. We could track it as `UpDownCounter`: we could increment it when enqueueing an item and decrement when dequeuing. But if we use `ObservableUpDownCounter`, we'll achieve the same more frugally by only returning the queue length every few seconds.

In a more complicated case of a distributed queue, we might not be able to instrument both the producer and consumer and would need to periodically get the current distributed queue length with a network call to the broker (be careful if you decide to do this in the counter callback).

Let's implement the queue length counter. First, the `Processor` class is disposable, so we should assume it can die before the application ends. It's important to disable all the instruments when this happens – we need to create a `Meter` as an instance variable and create the counter:

Processor.cs

```
_queueNameTag = new KeyValuePair<string, object?>("queue",
    queueName);
_meter = new Meter("queue.processor");
_queueLengthCounter = _meter
  .CreateObservableUpDownCounter(
    "queue.length",
    () => new Measurement<int>(queue.Count, _queueNameTag),
    "{items}",
    "Queue length");
```

https://github.com/PacktPublishing/Modern-Distributed-Trac-
ing-in-.NET/blob/main/chapter7/metrics/Processor.cs

Here, we created an `ObservableUpDownCounter` instance with a `queue.length` name and configured it to report the length in a callback, along with the queue name attribute. The last thing we need to do is to dispose of the `Meter` instance, along with the processor, using the `_meter.Dispose()` method. That's it!

Start the sample application (unless it's still running) with `metrics$ docker-compose up --build` and check out the `queue_length` metric (at `http://localhost:8889/metrics`). You should see it, among other metrics:

```
# HELP queue_length Queue length
# TYPE queue_length gauge
queue_length{job="metrics",queue="add"} 2
queue_length{job="metrics",queue="remove"} 0
queue_length{job="metrics",queue="update"} 157
```

As you can see, `UpDownCounter` and `ObservableUpDownCounter` are both mapped to gauge Prometheus – we'll learn more about gauges in the next section.

We can visualize this metric in the Prometheus UI (at `http://localhost:9090`) with the `avg by (queue) (queue_length)` query, as shown in *Figure 7.2*:

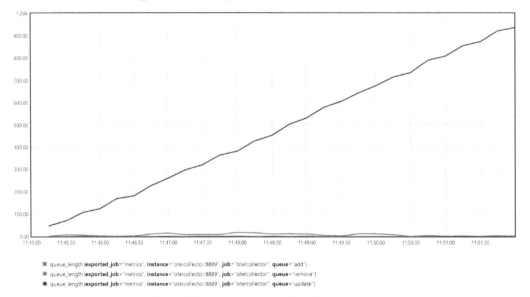

Figure 7.2 – Average queue length per queue

By looking at this chart, we can say that the **update** queue grows linearly while the other queues remain almost empty. We don't need any complicated queries here because we're interested in absolute values as we expect the queue length to always be small.

Let's learn about other instruments – gauges and histograms – and investigate what happens with the **update** queue.

Using an asynchronous gauge

`System.Diagnostics.Metrics.ObservableGauge` represents the current value of a non-additive metric. There is only an asynchronous version of it.

The key difference with `ObservableUpdownCounter` is that the counter is additive. For example, with counters, if we have multiple metric points for the same counter name, at the same timestamp, and with the same attributes, we can just add them up. For gauge, aggregation makes no sense and OpenTelemetry uses the last reported value.

When exported to Prometheus, `ObservableGauge` and `ObservableUpdownCounter` are the same, but their OTLP definitions (over-the-wire formats) are different.

> **Tip**
> You can get an idea of the internal representation of metric points on the OpenTelemetry side by enabling the `ConsoleExporter` output or looking into the OpenTelemetry documentation at `https://opentelemetry.io/docs/reference/specification/overview/#metrics-data-model-and-sdk`.

We use `ObservableGauge` to report a sequence number for the last processed item. It's useful with distributed queues, where the sequence number (or offset) represents the unique and ordered position of the item in the queue.

By looking at sequence number trends, we can estimate how many items are processed and how fast they are processed. For example, if the processor is stuck trying to process an invalid work item, we'd see that the sequence number is not increasing.

Adding up sequence numbers from different queues make no sense, so it should be an `ObservableGauge`, which we can create using a familiar API:

Processor.cs

```
_sequenceNumberGauge = _meter
  .CreateObservableGauge(
    "processor.last_sequence_number",
    () => new Measurement<long>(_seqNo, _queueNameTag),
    null,
    "Sequence number of the last dequeued item");
```

`https://github.com/PacktPublishing/Modern-Distributed-Tracing-in-.NET/blob/main/chapter7/metrics/Processor.cs`

In the callback, we return the `_seqNo` instance variable, which we update after dequeuing the work item. The only thing we need here is thread safety; we don't need precision since data is collected periodically.

We can report values with zero or more attributes or multiple measurements at once, so long as they have different attributes.

If we run the sample application with `metrics$ docker-compose up --build`, we can check the sequence number in Prometheus with a query such as `delta(processor_last_sequence_number[1m])`. It returns the delta per minute and is shown in *Figure 7.3*:

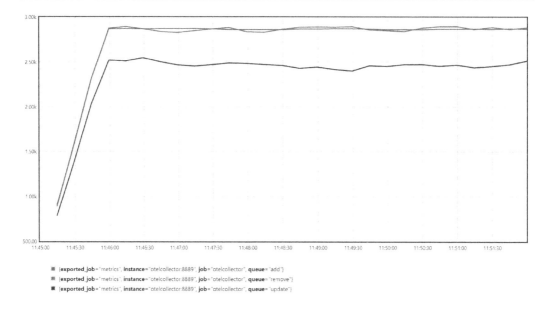

Figure 7.3 – Sequence number delta per minute

As we can see, after the application starts, the delta stabilizes around 3,000 items per minute in the **add** and **remove** queues, and 2,500 items in the **update** queue. This correlates with what we saw with the `queue_length` counter – the **update** queue is not processed fast enough. By looking at metrics, we can't say why, but there is one that can cast some light on it – the processing duration. Let's take a look at it.

Using histograms

`System.Diagnostics.Metrics.Histogram` represents a distribution of values – for example, the operation duration or payload size. Histograms can only be reported synchronously as each measurement is important. As we saw in *Chapter 2, Native Monitoring in .NET*, they allow us to calculate percentiles at query time.

We'll use a histogram to record the processing duration in our example:

Processor.cs

```
_processingDurationHistogram = _meter
  .CreateHistogram<double>(
    "processor.processing.duration",
    "ms",
    "Item processing duration");
```

https://github.com/PacktPublishing/Modern-Distributed-Trac-ing-in-.NET/blob/main/chapter7/metrics/Processor.cs

Every time we process an item from the queue, we should measure and record the time it took:

Processor.cs

```
Stopwatch? duration = _processingDurationHistogram
  .Enabled ? Stopwatch.StartNew() : null;
var status = await Process(item);
if (duration != null)
  _processingDurationHistogram.Record(
    duration.Elapsed.TotalMilliseconds,
    _queueNameTag,
    new KeyValuePair<string, object?>("status",
      StatusToString(status)));
```

https://github.com/PacktPublishing/Modern-Distributed-Trac-ing-in-.NET/blob/main/chapter7/metrics/Processor.cs

Here, we are using the Enabled flag – when metrics are not enabled, it prevents us from allocating a Stopwatch object on the heap.

The recording method has multiple overloads to report zero or more attributes associated with this value. Here, we report the queue name and the processing status. The status has low cardinality – it's an enum with a few values.

We also want to stay as efficient as possible, so we implemented the optimal and non-allocating StatusToString method.

Let's run the application with metrics$ docker-compose up --build and check out how the histogram looks in Prometheus exposition format (at http://localhost:8889/metrics). You should see a set of processor_processing_duration_milliseconds_bucket points for each queue, status, and bucket.

For example, this is what I see for the **add** queue with a status of Ok (attributes and some buckets have been omitted for brevity):

```
processor_processing_duration_milliseconds_bucket{le="0"} 0
...
processor_processing_duration_milliseconds_bucket{le="50"} 27
processor_processing_duration_milliseconds_bucket{le="75"} 52
processor_processing_duration_milliseconds_bucket{le="100"} 67
processor_processing_duration_milliseconds_bucket{le="250"} 72
...
```

```
processor_processing_duration_milliseconds_bucket{le="+Inf"} 72
processor_processing_duration_milliseconds_sum
  4145.833300000001
processor_processing_duration_milliseconds_count 72
```

Each bucket is identified by a `le` attribute – the inclusive upper boundary. There were 27 measurements smaller than or equal to 50 milliseconds, 52 measurements that were smaller than 75 milliseconds, and so on. Overall, there were 72 measurements, and the sum of all durations was around 4,146 milliseconds.

OTLP format defines a few more interesting properties that we can't see here:

- The `min` and `max` values for each bucket – they are not supported by Prometheus but show up in OTLP data.

- **Exemplars**, which represent examples of traces in a bucket. We could use them to easily navigate from metrics to traces and investigate long processing operations in higher histogram buckets. They are not implemented in OpenTelemetry for .NET yet.

The bucket boundaries we can see here are the default ones. They are static and work best if the measured value is well within the [0, 10000] range. If we start to measure values in the [10,000, 20,000] range, every measurement would be in the last two buckets, making the percentile calculation invalid. In this case, we should set explicit boundaries for corresponding histograms with the `MeterProviderBuilder.AddView` method.

In the future, OpenTelemetry will allow us to use exponential histograms with dynamic boundaries adjusted to data.

Note that we also have the `processor_processing_duration_milliseconds_sum` and `processor_processing_duration_milliseconds_count` metrics, so by reporting only the histogram, we get percentiles, averages, and measurement counters.

We can get the median processing time with the following query:

```
histogram_quantile(0.5, sum(rate(processor_processing_duration_
milliseconds_bucket{status="Ok"}[1m])) by (le, queue))
```

This should produce the chart shown in *Figure 7.4*:

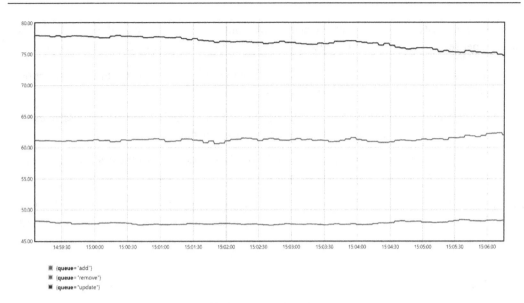

Figure 7.4 – Median processing time per queue

Here, we can see that the median processing time for the **add** queue is around 61 milliseconds, 48 milliseconds for the **remove** queue, and 75 milliseconds for the **update** queue.

Let's also check the processing rate using the sum by (queue) (rate(processor_processing_ duration_milliseconds_count[1m])) query, as shown in *Figure 7.5*:

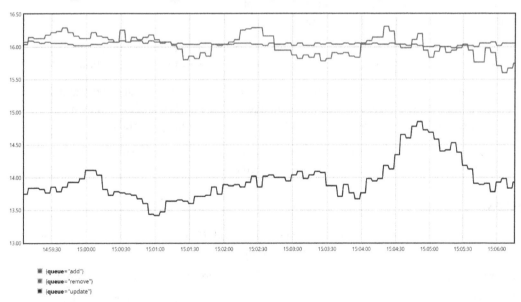

Figure 7.5 – Processing rate per queue

Items from the **update** queue are processed at a rate of about 14 items per second; the enqueue rate is ~16 items per second, as we saw in *Figure 7.1*. This should explain why the **update** queue is growing:

- Processing takes too much time – we should try to optimize it so that it targets at least 60 milliseconds to be able to process 16 items per second.

- If optimization is not possible (or is not enough), we know that we need to process an extra 2-3 items per second, so we need ~20% more processor instances.

- We could also implement backpressure on the producer side and throttle **update** requests to decrease the enqueue rate on the processor.

With just a small set of metrics, we were able to narrow down the problem to a specific area. If it was a production incident, we'd be able to quickly mitigate it by scaling the number of processors up and then investigating other options.

Summary

In this chapter, we explored metrics in .NET and OpenTelemetry.

Metrics allow us to collect aggregated multi-dimensional data. They produce unbiased telemetry with a predictable volume at any scale and allow us to monitor system health, performance, and usage.

Metrics can't have high-cardinality attributes, so we can't use them to detect problems that happen in specific and narrow cases – for this, we need distributed tracing or events. .NET provides an OpenTelemetry metrics implementation that consists of the `Meter` class, which can create specific instruments: counters, gauges, and histograms.

Counters are used to report additive values and can be synchronous or asynchronous. Gauges report current, non-additive values asynchronously, while histograms report value distribution.

With this, you should be able to identify scenarios where metrics are beneficial, choose appropriate instruments, and efficiently report metrics in your application. You should also be able to configure OpenTelemetry, and, most importantly, start detecting and monitoring performance issues.

In the next chapter, we're going to look at structured logs and events and learn how to write and consume them efficiently using .NET and OpenTelemetry.

Questions

1. Let's say you want to track the number of meme downloads (from our meme sample applications). Which telemetry signals would you choose? Why?

2. When reporting HTTP request duration, would you report it as a span, metric, or both?

3. How would you monitor the number of active application instances and the uptime?

Writing Structured and Correlated Logs

Distributed tracing is a great tool to describe and correlate operations, but sometimes, we need to record things such as callbacks and startup configurations, or conditionally write debug information. In this chapter, we're going to explore logs – the oldest and most popular telemetry signal that can describe anything.

First, we'll talk about logging use cases and discover different APIs available in .NET, and then we'll focus on `ILogger` – a common logging façade. We'll learn how to use it efficiently to write structured events. We'll see how to export logs with OpenTelemetry and write rich queries over them. Finally, we'll explore log sampling and cost-saving strategies.

In this chapter, you'll learn the following:

- When to write logs and which .NET API to use

- How to write logs with the `Microsoft.Extentions.Logging.ILogger` class

- How to capture and export logs with OpenTelemetry

- Cost-management strategies with the OpenTelemetry Collector

By the end of this chapter, you will be able to efficiently instrument your application with logs and events to debug and analyze service behavior.

Technical requirements

The code for this chapter is available in the book's repository on GitHub at `https://github.com/PacktPublishing/Modern-Distributed-Tracing-in-.NET/tree/main/chapter8`.

To run samples and perform analysis, we'll need the following tools:

- .NET SDK 7.0 or later

- Docker and `docker-compose`

Logging evolution in .NET

Logs are the most flexible telemetry signal and usually include a timestamp, a level, a category, a message, and sometimes attributes.

Logs are frequently intended to be human-readable and don't have a strict structure. Here's an example of a log record written to `stdout` by an ASP.NET Core application:

```
info: Microsoft.Hosting.Lifetime[14]
      Now listening on: http://localhost:5050
```

If we need to investigate something, we'd first look for logs describing interesting operations and then read the filtered logs. Our ability to understand what happened depends on how much context is logged and how searchable it is, with tools such as `grep`.

Structured logs are sometimes called **events**. Events are intended to be queried, potentially across multiple requests and based on any property, and need a strict and consistent structure. Here's the previous log record in the OpenTelemetry JSON format:

```
"timeUnixNano":"1673832588236902400",
"severityNumber":9, "severityText":"Information",
"body":{"stringValue":"Now listening on: {address}"},
"attributes":[
  {"key":"dotnet.ilogger.category",
     "value":{"stringValue":"Microsoft.Hosting.Lifetime"}},
  {"key":"Id","value":{"intValue":"14"},
  {"key":"address",
     "value":{"stringValue":"http://[::]:5050"}}], "traceId":"",
        "spanId":""}
```

It's not human-readable, but even when written to `stdout` or a file, it can be easily parsed into structured records without any prior knowledge of the semantics of the event.

As we started exploring in *Chapter 1*, *Observability Needs of Modern Applications*, the difference between logs and events is semantical – the same information can be optimized and printed in human-readable format or stored and indexed in a database in a structured format.

We're going to learn how to write such structured logs with the `Microsoft.Extensions.Logging.ILogger` class, but first, let's take a quick look at other logging APIs in .NET.

Console

We can use the `System.Console` class as a logger and write everything to `stdout`. We'd need to implement all logging primitives from scratch and forward `stdout` to the log management system, parsing it along the way to bring the original log structure back. Logging to `Console` is neither a convenient nor an efficient solution.

Trace

The `System.Diagnostics.Trace` and `System.Diagnostics.TraceSource` classes provide methods to write messages, along with some arguments, and support logging levels. We can also listen to them with the `TraceListener` class to export them to a log management system.

It seems like a good start, but there are a couple of limitations:

- The `TraceSource` API does not provide a standard way to write arguments. So, it's easy to format a message as a string, but we need to know specific event semantics to know argument names.

- By default, `TraceSource` and `TraceListener` use a global lock on every operation. It's possible to use them in a lock-free way, but it might be easy to overlook until the load becomes high enough.

So, `Trace` APIs solve some logging problems but introduce new ones.

EventSource

`System.Diagnostics.Tracing.EventSource` is another logging API in .NET. It's designed for high-performance scenarios, supports logging levels, and rich payloads, and captures the names and values of arguments. It's possible to listen to it by implementing the `EventListener` class or with .NET diagnostics tools running as a side-car process.

`EventSource` is a part of the .NET platform and can be used directly without any extra dependencies. `EventSource` is a perfect candidate to log in libraries that don't want to add any new dependencies.

When it comes to consumption, many observability vendors provide custom packages to listen to event sources, but there is no integration with OpenTelemetry yet, which is likely to change by the time you read it.

EventSource events can also be captured with `dotnet` diagnostics tools – `dotnet-trace` and `dotnet-monitor` – as we saw in *Chapter 4, Low-Level Performance Analysis with Diagnostic Tools,* and *Chapter 2, Native Monitoring in .NET.*

ILogger

`Microsoft.Extensions.Logging.ILogger` is a common logging façade integrated with ASP.NET Core. It supports structured logging and levels and has a rich ecosystem, making it easy to configure and send data to any provider, local or remote.

Logs written with `ILogger` can be consumed from other logging libraries, such as `Serilog` or `NLog`, and it's also supported by OpenTelemetry. Many observability backends support `ILogger`, making it a perfect tool to write application logs with.

`ILogger` logs can be captured out of process with .NET diagnostics tools. This is done by forwarding logs to `EventSource` first with the `Microsoft.Extensions.Logging.EventSource.EventSourceLoggingProvider` class. This provider is enabled by default in ASP.NET Core applications, and you can configure it manually with the `AddEventSourceLogger` extension method for the `ILoggingBuilder` interface. We used this mechanism to capture logs with `dotnet-monitor` and control log verbosity dynamically in *Chapter 2, Native Monitoring in .NET*.

Let's go through the `ILogger` usage in more detail.

Logging with ILogger

The `ILogger` class is part of the `Microsoft.Extensions.Logging.Abstractions` NuGet package. If you work on an ASP.NET Core application, worker service, or use other `Microsoft.Extensions` packages, you already depend on it transitively.

The `ILogger` interface exposes a few methods:

- `Log` records a log message with a given level, ID, exception, state, and formatter. The state type is generic but should contain a message, along with all the parameters and their names.

- `IsEnabled` checks whether logging at this level is enabled.

- `BeginScope` adds an object to the logging scope, allowing you to enrich nested log records with it. We saw scopes in action in *Chapter 2, Native Monitoring in .NET*, where we annotated console logs with trace context and ASP.NET Core request information.

It's common to use convenient extension methods defined in the `Microsoft.Extensions.Logging.LoggerExtensions` class instead of the vanilla `ILogger.Log` method.

Before writing any logs, let's first obtain an instance of `ILogger` – in ASP.NET Core applications, we can do it with constructor parameter injection, as shown in this example:

frontend/RetryHandler.cs

```
private readonly ILogger<RetryHandler> _logger;
public RetryHandler(ILogger<RetryHandler> logger) =>
    _logger = logger;
```

```
https://github.com/PacktPublishing/Modern-Distributed-Trac-
ing-in-.NET/blob/main/chapter8/memes/frontend/RetryHandler.cs
```

Here, we obtain an `ILogger` instance with the `RetryHandler` type parameter. The full name of the type parameter translates into the logging **category**, which is important to control verbosity and query logs, as we'll see later in the *Capturing logs with OpenTelemetry* section.

> **Note**
>
> Please refer to the .NET documentation at `https://learn.microsoft.com/aspnet/core/fundamentals/logging` to learn how to create and configure loggers.

Now, we can finally log things. For example, we can write an information log with `_logger.LogInformation("hello world")`.

If you use standard logging implementation, this call is broadcast to all registered logging providers that have the `Information` level enabled for this logging category.

Filters are provided at configuration time and can be global or specific to the logging provider. For example, here's a global logging configuration in our memes application:

frontend/appsettings.json

```
"Logging": {
  "LogLevel": {
    "frontend": "Information",
    "Microsoft.Hosting.Lifetime": "Information",
    "Default": "Warning"
  }
}
```

```
https://github.com/PacktPublishing/Modern-Distributed-Trac-
ing-in-.NET/blob/main/chapter8/memes/frontend/appsettings.json
```

This global configuration sets the `Information` level for the `frontend` and `Microsoft.Hosting.Lifetime` categories and `Warning` for everything else.

Let's get back to the `ILogger` API and see how we can write more useful logs. For example, let's log debug messages for error responses that include a response body.

We should be cautious here – a body stream usually can only be read once and can be very long, but in any case, we should be able to control any overhead that is introduced:

frontend/RetryHandler.cs

```
if (!response.IsSuccessStatusCode &&
    _logger.IsEnabled(LogLevel.Debug))
  _logger.LogDebug("got response: {status} {body} {url}",
    (int)response.StatusCode,
    await response.Content.ReadAsStringAsync(),
    response.RequestMessage?.RequestUri);
}
```

https://github.com/PacktPublishing/Modern-Distributed-Trac-ing-in-.NET/blob/main/chapter8/memes/frontend/RetryHandler.cs

Here, we write the log record at the Debug level and check whether the level is enabled *before* reading the response stream. We also use **semantic** (aka structured) logging, providing parameter names in curly brackets in the message string and their values as arguments.

> **Note**
>
> Make sure to use semantic logging. String interpolation or explicit formatting for ILogger messages removes the structure and makes performance optimization based on logging level impossible.

Arguments are passed as objects. ILogger implementations, such as OpenTelemetryLogger, support some types and usually call the ToString method on everything else. If logging at this level is not enabled, ToString is never called, saving you some CPU cycles and memory allocations.

Guarding logging calls, along with the retrieval or computation of arguments, with an IsEnabled check, is a great way to keep the performance impact of disabled categories very low.

Optimizing logging

Logging-related code frequently becomes a source of performance degradation. Avoiding memory allocations and the computation of arguments, especially when logging at this level is disabled, is the first step, but we should also optimize logging on the hot path when it's enabled. Here're a few tips for it:

- **Avoid excessive logging**: You might need to write a log record when entering an important code branch, a callback is called, or an exception is caught. Avoid logging exceptions multiple times as they propagate, or logging the same callback in nested methods.

- **Avoid duplication**: Unify multiple logs related to the same operation, and use logs coming from ASP.NET Core and other libraries when they are available, instead of adding your own.

- **Avoid calculating any values for logging purposes only**: It's common to serialize objects and parse or format strings, but this can usually be optimized by reusing existing objects, caching values, or formatting text at query time.

Finally, when log volume and arguments are optimized, we can do some micro-optimizations. One of them uses compile-time logging source generation and is demonstrated in the following example:

StorageService.cs

```
[LoggerMessage(EventId = 1, Level = LogLevel.Information,
  Message = "download {memeSize} {memeName}")]
private partial void DownloadMemeEvent(long? memeSize,
  string memeName);
```

https://github.com/PacktPublishing/Modern-Distributed-Trac-ing-in-.NET/blob/main/chapter8/memes/frontend/StorageService.cs

Here, we defined a partial method and annotated it with the `LoggerMessage` attribute, providing an event ID, level, and message. The implementation of this method is generated at compile time (you can find more information on it in the .NET documentation at https://learn.microsoft.com/dotnet/core/extensions/logger-message-generator).

If we check the generated code, we can see that it caches logger calls along with their static arguments. Please refer to the .NET documentation available at https://learn.microsoft.com/dotnet/core/extensions/high-performance-logging for more details on this approach.

We can compare the performance of different logging approaches by running `logging-benchmark$ dotnet run -c Release` and checking the results in the `BenchmarkDotNet.Artifacts` folder. The benchmark uses a dummy logger and measures the instrumentation side only. If we compare results for compile-time logging source generation and the `LogInformation` (or similar) method, we'll see the following results:

- Compile-time logging source generation eliminates memory allocations on the instrumentation side, even when logging is enabled. As a result, GC becomes less frequent, leading to higher throughput and smaller P95 latency.

- With compile-time logging source generation, an `IsEnabled` check is not needed if the argument values are readily available.

- The duration of an individual log call, when logging is enabled, does not depend much on the approach used.

These results may vary, depending on the argument types and values. Make sure to run performance, stress, and load tests, or profile your application with a similar logging configuration as used in production.

Now, you're fully equipped to write logs, so it's time to explore the consumption side.

Capturing logs with OpenTelemetry

By default, ASP.NET Core applications write logs to `stdout`, but since we want to correlate them with traces and query them by any attribute, we should export them to the observability backend or the log management tool that supports it. If your vendor supports `ILogger`, you can send logs directly to your vendor by configuring the corresponding logging provider. It will be up to this logging provider to annotate logs with trace context or environment information. By collecting logs with OpenTelemetry, we can annotate them consistently with other signals.

Let's see how to collect logs from the meme application with OpenTelemetry. To get the most out of the structure, we'll export them to **ClickHouse** – an open source database that supports SQL queries.

Here's an example of a configuration that exports logs with the **OpenTelemetry Protocol** (OTLP) exporter to the OpenTelemetry Collector first:

---

frontend/Program.cs

```
builder.Logging.AddOpenTelemetry(b => {
  b.SetResourceBuilder(resource);
  b.ParseStateValues = true;
  b.AddOtlpExporter();
});
```

https://github.com/PacktPublishing/Modern-Distributed-Tracing-in-.NET/blob/main/chapter8/memes/frontend/Program.cs

---

Here, we added an OpenTelemetry logging provider to the application's `ILoggingBuilder` instance and then configured the provider. We configured resource attributes, enabled parsing state values to populate arguments, and added the OTLP exporter. The exporter endpoint is configured with the `OTEL_EXPORTER_OTLP_ENDPOINT` environment variable.

The OpenTelemetry Collector is configured to send all logs to a file and write sampled logs to ClickHouse – we'll look into its configuration in the next section.

Let's go ahead and run the memes application with `memes$ docker-compose up --build`. Then, we'll hit the frontend at `http://localhost:5051/` to upload and download some memes.

To query logs in ClickHouse, run `$ docker exec -it memes-clickhouse-1 /usr/bin/clickhouse-client` – this will start a client where we can write SQL queries, such as the following one, that return all log records:

```
$ select * from otel_logs order by Timestamp desc
```

Here's an example of the output – the download meme log we added earlier in this chapter (if you don't see it, keep in mind that logs are sampled and you might need to download more memes):

```
| 2023-01-17 03:28:37.446217500 | 1bf63116f826fcc34a1e255
4b633580e | 2a6bbdfee21d260a |                1 | Information | 9
| frontend | download {memeSize} {memeName}|
{'service.instance.id':'833fb55a4717','service.name':'front
end'} | {'dotnet.ilogger.category':'frontend
.StorageService',
'Id':'1','Name':'DownloadMemeEvent',
'memeSize':'65412', 'memeName':'this is fine'}
```

It's barely readable but easy to query, as it includes a timestamp, a trace context, a log level, a body, resource information, and attributes – an event name, an ID, a meme size, and a name.

> **Note**
>
> At the time of writing, OpenTelemetry log specification is still experimental, so .NET implementation is minimal and details might change; the ClickHouse exporter is in alpha status, and the table schema could change in later versions.

We didn't enable capturing logging scopes; otherwise, we'd also see a few of them as attributes. They're populated by ASP.NET Core and describe incoming HTTP request properties. As we saw in *Chapter 2, Native Monitoring in .NET*, scopes include trace-context, which OpenTelemetry captures for us anyway.

With this, we can correlate logs using trace context or any attributes. For example, we can find the most popular memes with a query such as this:

```
select LogAttributes['memeName'], count(*) as downloads
from otel_logs
where ServiceName='frontend' and
  LogAttributes['Name']='DownloadMemeEvent'
group by LogAttributes['memeName'] order by downloads desc
limit 3
```

This can be useful when making business or technical decisions. For example, it helps to optimize caching or partitioning strategy, or plan capacity.

We can write queries such as these because we have enough structure in our logs, including an optional event ID and name. If we didn't have them, we'd have to filter logs based on message text, which is neither efficient nor reliable. For example, when someone changes the message when fixing a typo or adding new arguments, all saved queries need to be changed to reflect this.

> **Tip**
> To make logs queryable, make sure to use semantic logging. Provide a static event ID and name. Use consistent (across the whole system) attribute names.

By following this approach, we can change observability vendors, print logs in human-readable format, and, at the same time, store them in a structured form, post-process them, or aggregate them if needed.

Structured logs combined with traces allow us to report business telemetry and run queries, but it brings new costs – let's see how we can control them.

Managing logging costs

Similarly to tracing and metrics, logging increases the compute resources needed to run an application, the cost of running a logging pipeline (if any), and the costs associated with using (or running) an observability backend. Vendor pricing is frequently based on a combination of telemetry volume, retention time, and API calls, including queries.

We already know how to write logs efficiently, so let's talk about pipelines and backends.

Pipelines

A logging pipeline consists of the infrastructure needed to send logs to the backend of your choice. It's typical to do some grokking, parsing, transformations, buffering, throttling, and hardening on the way to the backend.

In a simple case, it's all done by your vendor's logging provider or the OpenTelemetry processors and exporter inside the process.

In many cases, we need logging pipelines to capture logs and events coming from outside – the OS, self-hosted third-party services, proxies, and other infrastructure pieces. They could be structured such as Kubernetes events, have a well-known configurable format such as HTTP server logs, or have no structure at all.

A logging pipeline can help parse such logs and transform them into a common format. In the OpenTelemetry world, this could be done on the Collector.

We would receive logs from `files`, `syslog`, `journald`, `fluentd`, other systems, or collectors with a **receiver**, then massage, filter, and route them with a **processor**, and finally, export them to the final destination.

Cost-saving strategies for pipelines start with a typical approach to minimize log volume and avoid duplication and complex transformations, as we discussed earlier in this chapter.

For example, you might enable HTTP tracing, metrics, and logs from both a client and server and logs from the HTTP proxy as well. Do you need logs from the proxy? Do you use them?

Eliminate duplicates by potentially substituting them with metrics, less verbose events, or attributes on other signals. If some information is rarely needed, process it lazily.

It's also important to monitor your logging pipeline – measure the error rate and estimate the end-to-end latency and throughput. The OpenTelemetry Collector helps by exposing its own metrics and logs.

Once, the team I worked on discovered that some logs had dropped at an ~80% rate within the logging pipeline. We published them in a fire-and-forget manner and didn't even know they were dropped until we were not able to investigate incidents in production.

Backends

Backend cost optimization also starts by producing as few logs as possible. Then, costs can be controlled in different ways, depending on your constraints and observability backend pricing model:

- The log volume can be reduced with sampling. Aggregations based on sampled events would need to be scaled accordingly but would provide unskewed results when unbiased sampling is used. Logs can be sampled consistently with traces at the same or a higher rate.

- Logs can stay in hot storage for a short period of time and then move to cold storage. During the first few days, logs in hot storage can be used for urgent ad hoc queries, but after that, query speed becomes less important.

  This strategy can be combined with sampling – logs could be sent to the cold (and cheap) storage, while sampled in logs would go to hot storage.

- Certain logs can be post-processed and aggregated into metrics or reports for frequent queries.

All these strategies and combinations of them can be implemented with the OpenTelemetry Collector. For example, in our memes application, we use a combination of sampling and hot/cold storage, as shown in *Figure 8.1*:

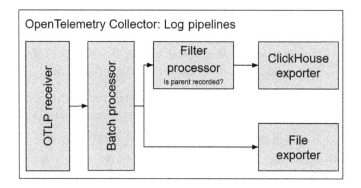

Figure 8.1 – Logging pipelines with sampling and hot and cold storage

We have two different logging pipelines here. Both start with the OTLP receiver and batch processor. Then, one pipeline writes all logs to a file, and another one runs a filter based on log record properties. It checks trace-flags and drops logs when the parent span is not recorded. Logs with a recorded parent (or those that have no parent at all, such as startup logs) end up in ClickHouse. Here's the corresponding logging pipeline configuration:

otel-collector-config.yml

```
logs:
  receivers: [otlp]
  processors: [batch]
  exporters: [file]
logs/sampled:
  receivers: [otlp]
  processors: [batch, filter]
  exporters: [clickhouse]
```

https://github.com/PacktPublishing/Modern-Distributed-Trac-ing-in-.NET/blob/main/chapter8/memes/configs/otel-collector-con-fig.yml

The filter processor, and many other processors, leverage a rich transformation language – **OTTL**. OTTL can be used to rename attributes, change their values, drop metrics and spans, create derived metrics, or add and drop attributes. Here's the filter processor configuration:

```
filter:
  logs:
    log_record:
      - 'flags == 0 and trace_id != TraceID
        (0x00000000000000000000000000000000)'
```

https://github.com/PacktPublishing/Modern-Distributed-Trac-ing-in-.NET/blob/main/chapter8/memes/configs/otel-collector-con-fig.yml

The collector can solve many common post-processing needs and take this burden away from your service.

That brings us to the end of this chapter. Let's recollect what we've learned so far.

Summary

Logs are the most flexible telemetry signal – they can be used to write information in human-readable format, complement traces with more information, or record structured events to analyze usage or performance.

To write logs, we can use different logging APIs – `ILogger` works best for application code, while `EventSource` is usually the best choice for libraries.

`ILogger` makes it easy to write structured logs efficiently, but it depends on application authors to do so by minimizing log volume and the operations needed to calculate logging arguments.

`ILogger` has a rich ecosystem of integrations with .NET frameworks, libraries, and providers that can send logs almost anywhere in a flat or structured format.

Collecting and exporting `ILogger` logs with OpenTelemetry produces logs that are consistent and correlated with other telemetry signals.

In addition to application logs, we usually also need to collect logs from infrastructure or legacy systems. We can do it with the OpenTelemetry Collector, which allows us to collect and unify logs from multiple destinations. The Collector's logging pipelines can throttle, aggregate, or route logs to help you manage your logging costs.

You should now be ready to efficiently instrument your application with structured logs and export them with OpenTelemetry. You're also prepared to build logging pipelines with OpenTelemetry to add observability to your infrastructure and control logging costs.

This concludes our deep dive into individual telemetry signals. In the next chapter, we'll talk about choosing a good set of telemetry signals, depending on a scenario, and adding the right level of information, based on OpenTelemetry semantic conventions.

Questions

1. Is the following code snippet correct? How would you improve it?

    ```
    var foo = 42;
    var bar = "bar";
    logger.LogInformation($"hello world: {foo}, {bar}");
    ```

2. Let's say your application writes usage events using the ILogger APIs. Events are exported somewhere and then used to build business-critical reports. As your application evolves, you will probably refactor code, rename namespaces and classes, improve log messages, and add more arguments. How can you write logs to keep the usage report resilient to logging changes?

3. Assuming that traces for HTTP requests are collected, do you also need to write logs for the same HTTP calls?

Part 3: Observability for Common Cloud Scenarios

This part provides instrumentation recipes for common scenarios such as network calls, async messaging, databases, and web clients. It demonstrates how to write your own instrumentation or cover a gap in an automatic one, and, most importantly, how to investigate performance issues using a combination of distributed tracing, metrics, and logs.

This part has the following chapters:

- *Chapter 9, Best Practices*
- *Chapter 10, Tracing Network Calls*
- *Chapter 11, Instrumenting Messaging Scenarios*
- *Chapter 12, Instrumenting Database Calls*

9

Best Practices

In the previous chapters, we focused on how to collect, enrich, correlate, and use individual telemetry signals. In this chapter, we're going to discuss what information to collect and how to represent it efficiently using all the available signals. We'll start by providing recommendations on how to pick a suitable telemetry signal and suggest cross-signal cost optimization strategies. Finally, we'll explore about OpenTelemetry semantic conventions and use them to create consistent telemetry supported by most observability vendors.

You'll learn how to do the following:

- Find telemetry signals that work for your scenarios
- Control telemetry costs with aggregation, sampling, and verbosity
- Follow common practices when reporting telemetry with OpenTelemetry semantic conventions

By the end of this chapter, you will be able to use existing semantics for common technologies or create your own cross-signal and cross-service conventions.

Technical requirements

There are no specific requirements for this chapter, and there are no associated code files either.

Choosing the right signal

When we discussed individual telemetry signals in *Chapters 6* to *8*, we provided suggestions on when to use each of them. Let's do a quick recap:

- **Distributed traces** describe individual network calls and other interesting operations in detail. Spans have causal relationships, allowing us to understand the request flow in distributed systems.

  Traces document the request flow through the system and are essential for investigating errors or outliers in the long tail of latency distribution. Traces provide means to correlate other telemetry signals.

- **Metrics** collect aggregated data with low-cardinality attributes and provide a low-resolution view of the overall system state. They help optimize telemetry collection and reduce storage costs and query time.

- **Events** provide highly structured information about individual occurrences of important things. The key difference between spans and events is that spans have unique contexts and describe something that lasts.

  Events have high-cardinality attributes and can help answer ad hoc questions about system behavior and usage.

- **Logs** provide details about operations in a human-readable and less structured format.

  They are useful for debugging things when other signals don't provide enough information. Also, logs are the only signal that supports verbosity.

- **Profiles** are low-level performance data describing individual operations within a single process that helps optimize performance and identify resource bottlenecks.

When instrumenting some specific scenario, we usually need a combination of signals.

For example, to get observability into network calls, we need traces to ensure we can track the request flow across services and correlate other signals. Logs are necessary to record exceptions and warnings, describe local operations, and provide debug-level data for complicated investigations. Finally, we may need metrics to record non-sampled measurements, optimize collection, and reduce query time.

> **Note**
> Before thinking about signals, we should have an idea of what information we want to be available, how we're going to use it, how fast and frequently we need it, how many details we want to capture, for how long we need it, how much we can afford, and the downtime cost.

The answers to these questions should shape our decisions around observability.

Essentially, we have multiple trade-offs between having enough data to investigate issues fast and the cost of the observability solution. For example, collecting too many traces would give us all the details we need to investigate all sorts of issues. It would have a noticeable performance impact and significantly increase observability backend costs. As a result, traces might become so deep and detailed that it would be hard to understand where the problems are.

The conversation about a good set of telemetry signals is not possible without talking about costs. Let's see how we can control them.

Getting more with less

Since we usually need to collect multiple signals about the same component, we need to be able to tune them individually, depending on our needs.

The key is to reduce the volume of expensive, but not essential, data, potentially replacing it with cheaper options while keeping the system observable. We saw how we can do this by combining hot and cold storage or changing the retention period in *Chapter 8, Writing Structured and Correlated Logs*. Here, let's focus on the collection side.

While observability vendors have different pricing models, it's common for them to bill for traces, logs, and events depending on the volume, and for metrics depending on the number of time series. Queries (or API calls) can also be charged for and may have concurrency limits.

Of course, we can always add or remove instrumentations or stop writing logs and events, but there are a few more factors affecting how much telemetry is collected:

- We can control tracing volume with the sampling rate and by adding or removing new attributes
- To control the number of metric time series, we can add or remove resource attributes or drop dimensions or instruments
- We can tune logging verbosity for individual categories or do so globally and add or remove attributes

Applications' needs may vary, depending on their maturity, the number of changes, the downtime they can afford, and other factors – let's go through several examples to demonstrate possible compromises they can apply.

Building a new application

When writing the first version of an application, telemetry can play a critical role in helping teams investigate issues and move faster. The interesting part here is that we don't know which type of telemetry we need and how we're going to use it.

We can leverage existing instrumentations that allow us to focus our efforts on building the application and having all means to debug it as it evolves, while also finding answers to questions about telemetry we outlined before.

The initial stages are a great time to design the observability story and it makes sense to start with the most flexible signals – traces, events, and logs. Initially, telemetry volume is likely to be low, so recording traces with a high sampling rate or just rate-limiting should be affordable. Also, we probably don't have strict SLAs yet and don't use dashboards and alerts much.

Until we get some real users, metrics or events might be unnecessary, but this is a good time to experiment and get familiar with them.

Even if the telemetry volume is quite low and we can capture verbose data, we should avoid adding excessive amounts of telemetry and should remove unused signals.

Evolving applications

As our application starts getting some real users, getting performance data quickly becomes critical. By this time, we have more clarity on what's important to measure in the application and how to debug issues (hopefully not relying on verbose logging).

This is the time to optimize and tune telemetry collection. As the load grows, we usually want to lower the sampling rate for traces and reduce log verbosity.

Also, we would probably need to create alerts and build dashboards that are much more efficient when done over metrics, as we discussed in *Chapter 7, Adding Custom Metrics*. While instrumentation libraries should cover the basics, we might need to add custom metrics where we previously relied on queries over traces. As we scale up, the number of time series only increases with the number of service instances.

At this stage, we might also decide to collect precise and unsampled usage data with events and metrics.

The application is still changing a lot and we frequently need to investigate functional issues for specific requests and optimize requests from the long tail of latency. So, tracing still plays a key role in day-to-day work. We might need to instrument more layers in the application to capture logical operations or add applications-specific context. At the same time, we may find some auto-instrumentations too verbose and can tune them to remove unnecessary attributes or suppress some spans.

Sometimes, we need to capture profiles or use diagnostic tools to investigate lower-level issues, so having a continuous profiler or adding `dotnet-monitor` in a sidecar could make such investigations much easier.

If the application (or some parts of it) becomes more stable due to having fewer and fewer issues, it makes sense to remove non-essential traces and reduce the sampling rate for stable services or endpoints. Tail-based sampling could help capture more traces for failures or long requests.

When the application is not changing anymore except for basic maintenance, but more importantly, if it does not have many issues and investigations, it could be reasonable to reduce tracing to just incoming and outgoing requests, potentially forwarding logs to colder storage.

Performance-sensitive scenarios

Instrumentation introduces performance overhead. Between traces, metrics, and logs, traces are the most expensive. When instrumenting an HTTP request, this overhead is usually negligible compared to the call itself.

But in some cases, instrumentation costs can be too high. For example, when returning cached responses or rate-limiting requests across all service instances, logging or tracing all such calls can significantly impact performance. Moreover, if we recorded a trace for every request, a DDOS attack or buggy client might kill our observability pipeline, if not the whole service.

Tracing overhead, to some extent, can be reduced with sampling, which protects the observability pipeline and reduces the number of allocations when populating attributes, but a new `Activity` is created and a new `SpanId` is generated, regardless of the sampling decision.

Adding tracing for a hot path should be done with caution. Keep the number of traces to a minimum: trace incoming requests only if the corresponding request is going to be processed by your application and avoid tracing outgoing network calls to the leaf services if they're extremely fast or reliable. For example, it makes sense to report an event instead of a span when talking to Redis.

Metrics are the most performant telemetry signal and should be preferred for a hot path when possible. For example, reporting the Redis call duration as a metric with a cache hit/miss dimension would likely be cheaper than an event. And for tracing purposes, we can put a hit/miss flag as an attribute on an existing current span (for example, one representing an incoming request).

Recording exceptions and errors is usually fine from a performance perspective since exceptions create a huge overhead anyway. But in the case of a failure storm, we get too many of them, so it's a good idea to throttle exception reporting.

Implementing efficient, useful, but minimalistic instrumentation usually requires several iterations. Luckily, OpenTelemetry provides a set of semantic conventions for common scenarios that can help with it. Let's see how.

Staying consistent with semantic conventions

One of the most important questions we're yet to discuss is what information to add to telemetry signals to make them useful – this is where OpenTelemetry semantic conventions come into play.

Semantic conventions describe what information to collect for specific technologies, such as HTTP or gRPC calls, database operations, messaging scenarios, serverless environments, runtime metrics, resource attributes, and so on.

Semantic conventions are part of the OpenTelemetry specification and have been published in the specification repository at `https://github.com/open-telemetry/opentelemetry-specification`. They apply to all instrumentations authored by the OpenTelemetry project.

> **Note**
> At the time of writing, semantic conventions are in an experimental status. The community is actively working on stabilization and the attributes I use in this book will likely be renamed or changed in other ways.

The goal of semantic conventions is to unify telemetry collection for specific scenarios or technology across languages, runtimes, and libraries. For example, traces and metrics for all HTTP clients look very similar, making it possible to visualize or query HTTP telemetry or diagnose problems in the same way for any application. Let's look at HTTP semantic conventions to understand how they work and give you an idea of what other conventions look like.

Semantic conventions for HTTP requests

The conventions cover tracing and metrics for incoming and outgoing HTTP requests. Spans with `client` kind describe outgoing requests, whereas `server` spans describe incoming requests. Instrumentations create a new span for each attempt.

`client` HTTP spans contain attributes that describe the request, response, and remote destination. According to the current version, a minimal HTTP client instrumentation must report the following attributes: `http.method`, `http.url`, `net.peer.name`, `net.peer.port`, and `http.status_code`.

If a response is not received, the `http.status_code` attribute is not populated; instead, the span status would indicate an error and provide a status description that explained what happened. The port (`net.peer.port`) attribute may be skipped if it's 80 or 443. Other attributes are required, so all instrumentations that follow conventions must populate them in all scenarios. These attributes, combined with the span start timestamp, duration, and status, provide a minimal necessary description of the HTTP request.

All the attributes except `http.status_code` should be provided at the span start time – this allows us to make sampling decisions based on these attributes.

You probably noticed that the host and port information is available inside the URL and via separate attributes. The URL is a high-cardinality attribute, but the host and port are very likely to be of low cardinality, so reporting all of them allows us to unify instrumentation code and report traces and metrics in one place. It also makes it possible to calculate metrics from traces and simplify queries.

Minimal HTTP server instrumentation reports the `http.method`, `http.status_code`, `http.scheme`, `http.target`, `net.host.name`, `net.host.port`, and `http.route` attributes.

Since HTTP servers don't have full URLs readily available, instrumentations don't construct them and report individual URL components instead. Route information is provided by an HTTP framework such as ASP.NET Core and even there, you may handle requests in middleware without using routing. Reporting route is quite important for metrics, as we've seen in *Chapter 7, Adding Custom Metrics*, so if you don't have the route available out of the box, you might want to provide one manually to distinguish different classes of API calls. HTTP client and server instrumentations usually also report recommended attributes, such as the `User-Agent` header, request and response content length, HTTP protocol version, and remote IP address.

Conventions also standardize attribute value types – for example, `http.status_code` has an integer type, simplifying comparison at query time.

You can find the full HTTP tracing conventions at `https://opentelemetry.io/docs/reference/specification/trace/semantic_conventions/http`.

Metrics are based on the same tracing attributes and cover request duration, content size, and the number of active requests on servers. The metrics conventions are available at `https://opentelemetry.io/docs/reference/specification/metrics/semantic_conventions/http-metrics`.

HTTP semantic conventions provide a good set of default things to collect. You can move between teams, companies, and web frameworks, or start using a different programming language, but OpenTelemetry instrumentations would provide a common baseline everywhere.

Having a reliable set of required attributes helps the backend visualize traces and service maps, build dashboards, and automate analysis and issue detection.

General considerations

When you need to instrument some specific technology or scenario and no suitable instrumentation library is available, make sure to also check whether there is an applicable semantic convention. By following it, you will be able to leverage any experiences built on top of it by your observability backend, prevent inconsistent signals coming from different parts of your system, and also save some time designing and polishing your signals.

But what if you want to instrument something very specific to your application, such as adding spans for logical operations or adding usage metrics? Let's see.

Tracing

As we've seen in *Chapter 6, Tracing Your Code*, we can create a new `Activity` instance without specifying any parameters. By default, it's named after the caller method and has an `internal` kind.

OpenTelemetry recommends using low-cardinality span names. HTTP client span names follow the `HTTP <method>` pattern (for example, `HTTP GET`), while the HTTP server span name looks like `<method> <route>` (for example, `GET /users/{userId}`). The span name describes a class of operations and is frequently used to group common spans.

Another important property is the span kind: it helps backends visualize and query traces them. `client` spans represent outgoing requests – their context is propagated over the wire, and they become remote parents of `server` spans. When instrumenting a remote call, we would typically want to create a new span for each attempt so that we know how long an attempt took, how many there were, and what the backoff interval was.

The `server` spans are those that track incoming requests; they either have no parents or have a remote parent.

OpenTelemetry also defines `consumer` and `producer` kinds – they are used in asynchronous scenarios where a request-response pattern is not applicable. A `producer` span could be a parent of a `consumer` span (or be linked to it), but it usually ends before the corresponding `consumer` span.

All other spans are `internal`. For example, to represent an I/O operation or a local long-running call, we should use the `internal` kind. When instrumenting client library calls or logical operations that can do multiple HTTP requests underneath, it makes sense to describe them as `internal` spans.

If an operation ends with an error, we should reflect it with a span status, but this can be tricky. For example, HTTP semantic conventions recommend setting the status to an error on the client side if a response was not received, there were too many redirects, or when the status code was in the 4xx or 5xx ranges. But for HTTP servers, a 4xx response does not indicate an error and should be left unset. Even for client requests, status codes such as 404 (`Not Found`) do not necessarily indicate an error and can be used to check whether some resource exists.

When recording errors, the status description can be used to record some predictable and short information about it, such as its exception type and/or message. Exceptions follow their own semantic conventions – we discussed this in *Chapter 6*, *Tracing Your Code*. They can be huge (because of stack traces), so we should avoid recording handled exceptions.

Attributes

Application-specific context or details about an operation can be recorded in attributes. Before inventing a new attribute name, make sure you check existing semantic conventions to see whether something similar is defined there already. For example, you can use general network attributes to describe remote destinations or host and RPC calls.

If you must create a new attribute, use a short name that consists of basic Latin characters. OpenTelemetry recommends using namespaces to avoid naming collisions (they are delimited with the dot (`.`) character) and using `snake_case` to separate words. For example, in `http.status_code`, `http` is a namespace. So, if you're defining a new attribute specific to your company, it makes sense to use the company name in the namespace.

The number of attributes per span is limited to 128 by default, but this limit can be increased.

Keeping consistent names and value types across your system can be challenging, so it's a good idea to come up with some registry to keep them consistent.

So, which information would you add to attributes? Anything that describes your operation, except sensitive information or secrets. Be cautious with long values and avoid adding something that needs to be serialized or calculated – use verbose logging for it.

It's also a good idea to avoid duplication and record a reasonable set of information, moving static attributes to resources instead of spans.

Metrics

When creating instruments, we can provide a name, unit, and description.

Instrument names are case-insensitive and consist of alphanumeric characters, underscores, dots, and dashes. Instrument names must be short – up to 63 characters.

Instrument names are formatted similarly to attribute names and support namespaces – for example, the `http.server.active_requests` counter or the `http.server.duration` histogram, which represent the number of active HTTP requests and server-side duration of requests, respectively.

Units usually follow UCUM standards (`https://ucum.org/`) and it's important to keep them consistent for the same instrument across the whole system.

Attribute naming conventions are common between different signals and usually, metrics rely on a subset of tracing attributes. The most important characteristic of metric attributes is low cardinality, which we described in *Chapter 7, Adding Custom Metrics*.

Before adding custom metrics, make sure to check whether there is an existing instrumentation library or an OpenTelemetry semantic convention. For example, there is a generic one for RPC requests, process and system resource utilization metrics, databases, and other technology-specific ones, such as Kafka.

Summary

In this chapter, we discussed suggestions and recommendations for telemetry collection. To describe some scenario or operation, we usually need multiple signals: tracing enables correlation and causation, logs provide additional information not covered by traces, events collect usage information, and metrics optimize instrumentations, queries, and alerts.

Depending on your application's needs and stability, you can control costs by tuning the sampling rate on tracing and using metrics for performance data and events for usage reports.

OpenTelemetry semantic conventions provide instrumentation recipes for common technologies and concepts. By following them, you can create high-quality instrumentations with good defaults that you can tune to your needs. Observability backends can provide their best experiences to help you visualize, detect anomalies, and perform other semi-automated analyses. For proprietary technologies or application-specific instrumentation, where there are no existing conventions, it's important to follow general the OpenTelemetry specification and naming patterns and report telemetry consistently across your system.

With this, you should be ready to instrument advanced scenarios with multiple signals and provide a rich context while following the available practices. In the next chapter, we're going to apply these skills to instrument gRPC streaming calls that are not covered by any existing conventions. Stay tuned.

Questions

1. Can you instrument a tiny stateless RESTful microservice with tracing only?

2. When working on an application that processes thousands of requests per second on each instance, which sampling rate would you choose?

3. Your application communicates with client devices over web sockets. How would you approach instrumenting this communication?

10

Tracing Network Calls

In this chapter, we'll apply what we learned about tracing in *Chapter 6, Tracing Your Code*, to instrument client and server communication via gRPC.

We'll start by instrumenting unary gRPC calls on the client and server according to OpenTelemetry semantic conventions. Then, we'll switch to streaming and explore different ways to get observability for individual messages. We'll see how to describe them with events or individual spans and learn how to propagate context within individual messages. Finally, we'll see how to use our instrumentation to investigate issues.

In this chapter, you'll learn how to do the following:

- Instrument network calls on the client and server following OpenTelemetry semantic conventions and propagate context over the wire

- Instrument gRPC streaming calls according to your application needs

- Apply telemetry to get insights into network call latency and failure rates and investigate issues

Using gRPC as an example, this chapter will show you how to trace network calls and propagate context through them. With this chapter, you should also be able to instrument advanced streaming scenarios and pick the appropriate observability signals and granularity for your traces.

Technical requirements

The code for this chapter is available in the book's repository on GitHub at `https://github.com/PacktPublishing/Modern-Distributed-Tracing-in-.NET/tree/main/chapter10`.

To run the samples and perform analysis, we'll need the following tools:

- .NET SDK 7.0 or later

- Docker and `docker-compose`

Instrumenting client calls

Network calls are probably the most important thing to instrument in any distributed application since network and downstream services are unreliable and complex resources. In order to understand how our application works and breaks, we need to know how the services we depend on perform.

Network-level metrics can help us measure essential things such as latency, error rate, throughput, and the number of active requests and connections. Tracing enables context propagation and helps us see how requests flow through the system. So, if you instrument your application at all, you should start with incoming and outgoing requests.

When instrumenting the client side of calls, we need to pick the right level of the network stack. Do we want to trace TCP packets? Can we? The answer depends, but distributed tracing is usually applied on the application layer of the network stack where protocols such as HTTP or AMQP live.

In the case of HTTP on .NET, we apply instrumentation on the `HttpClient` level – to be more precise, on the `HttpMessageHandler` level, which performs individual HTTP requests, so we trace individual retries and redirects.

If we instrument `HttpClient` methods, in many cases, we collect the duration of the request, which includes all attempts to get a response with back-off intervals between them. The error rate would show the rate without transient failures. This information is very useful, but it describes network-level calls very indirectly and heavily depends on the upstream service configuration and performance.

Usually, gRPC runs on top of HTTP/2 and to some extent can be covered by HTTP instrumentation. This is the case for unary calls, when a client sends a request and awaits a response. The key difference with HTTP instrumentation is that we'd want to collect a gRPC-specific set of attributes, which includes the service and method names as well as the gRPC status code.

However, gRPC also supports streaming when the client establishes a connection with the server, and then they can send each other multiple asynchronous messages within the scope of one HTTP/2 call. We'll talk about streaming calls later in the *Instrumenting streaming calls* section of this chapter. For now, let's focus on unary calls.

Instrumenting unary calls

We're going to use gRPC implementation in the `Grpc.Net.Client` NuGet package, which has an OpenTelemetry instrumentation library available.

> **Note**
> OpenTelemetry provides two flavors of gRPC instrumentation: one for the `Grpc.Net.Client` package called `OpenTelemetry.Instrumentation.GrpcNetClient` and another one for the lower-level `Grpc.Core.Api` package called `OpenTelemetry.Instrumentation.GrpcCore`. Depending on how you use gRPC, make sure to use one or another.

These instrumentations should cover most gRPC tracing needs and you can customize them further using the techniques described in *Chapter 5*, *Configuration and Control Plane*. For example, the `OpenTelemetry.Instrumentation.GrpcNetClient` instrumentation allows the suppression of the underlying HTTP instrumentation or the enrichment of corresponding activities.

Here, we're going to write our own instrumentation as a learning exercise, which you can apply to other protocols or use to satisfy additional requirements that you might have.

We can wrap every gRPC call with instrumentation code, but this would be hard to maintain and would pollute the application code. A better approach would be to implement instrumentation in a gRPC `Interceptor`.

So, we know where instrumentation should be done, but what should we instrument? Let's start with gRPC OpenTelemetry semantic conventions – the tracing conventions are available at `https://github.com/open-telemetry/opentelemetry-specification/blob/main/specification/trace/semantic_conventions/rpc.md`.

The conventions are currently experimental and some changes (such as attribute renames) should be expected.

For unary client calls, the tracing specification recommends using the `{package.service}/{method}` pattern for span names and the following set of essential attributes:

- The `rpc.system` attribute has to match `grpc` – it helps backends understand that it's a gRPC call.
- The `rpc.service` and `rpc.method` attributes should describe the gRPC service and method. Even though this information is available in the span name, individual service and method attributes help query and filter spans in a more reliable and efficient way.
- The `net.peer.name` and `net.peer.port` attributes describe remote endpoints.
- `rpc.grpc.status_code` describes the numeric representation of the gRPC status code.

So, in the interceptor, we need to do a few things: start a new `Activity` with the recommended name and a set of attributes, inject the context into the outgoing request, await the response, set the status, and end the activity. This is demonstrated in the following code snippet:

client/GrpcTracingInterceptor.cs

```
public override AsyncUnaryCall<Res>
  AsyncUnaryCall<Req, Res>(Req request,
    ClientInterceptorContext<Req, Res> ctx,
    AsyncUnaryCallContinuation<Req, Res> continuation)
{
  var activity = Source.StartActivity(ctx.Method.FullName,
    ActivityKind.Client);
  ctx = InjectTraceContext(activity, ctx);
  if (activity?.IsAllDataRequested != true)
    return continuation(request, ctx);

  SetRpcAttributes(activity, ctx.Method);

  var call = continuation(request, context);
  return new AsyncUnaryCall<Res>(
    HandleResponse(call.ResponseAsync, activity, call),
    call.ResponseHeadersAsync,
    call.GetStatus,
    call.GetTrailers,
    call.Dispose);
}
```

https://github.com/PacktPublishing/Modern-Distributed-Trac-ing-in-.NET/blob/main/chapter10/client/GrpcTracingInterceptor.cs

Here, we override the interceptor's `AsyncUnaryCall` method: we start a new `Activity` with the client kind and inject a trace context regardless of the sampling decision. If the activity is sampled out, we just return a continuation call and avoid any additional performance overhead.

If the activity is sampled in, we set the gRPC attributes and return the continuation call with the modified response task:

client/GrpcTracingInterceptor.cs

```
private async Task<Res> HandleResponse<Req, Res>(Task<Res>
  original, Activity act, AsyncUnaryCall<Req> call)
{
  try
  {
```

```
    var response = await original;
    SetStatus(act, call.GetStatus());
    return response;
  }
  ...
  finally
  {
    act.Dispose();
  }
}
```

https://github.com/PacktPublishing/Modern-Distributed-Trac-ing-in-.NET/blob/main/chapter10/client/GrpcTracingInterceptor.cs

We dispose of `Activity` explicitly here since the `AsyncUnaryCall` method is synchronous and will end before the request is complete, but we need the activity to last until we get the response from the server.

Let's take a closer look at each of the operations, starting with context injection:

client/GrpcTracingInterceptor.cs

```
private ClientInterceptorContext<Req, Res>
  InjectTraceContext<Req, Res>(Activity? act,
    ClientInterceptorContext<Req, Res> ctx)
  where Req: class where Res: class
{
  ...
  _propagator.Inject(new PropagationContext(
      act.Context, Baggage.Current),
    ctx.Options.Headers,
    static (headers, k, v) => headers.Add(k, v));
  return ctx;
}
```

https://github.com/PacktPublishing/Modern-Distributed-Trac-ing-in-.NET/blob/main/chapter10/client/GrpcTracingInterceptor.cs

Here, we inject the context using the `OpenTelemetry.Context.Propagation.TextMapPropagator` class and the `Inject` method. We'll see how the propagator is configured a bit later.

We created an instance of the `PropagationContext` structure – it contains everything that needs to be propagated, namely `ActivityContext` and the current `Baggage`.

The context is injected into the `ctx.Options.Headers` property, which represents gRPC metadata. The metadata is later on transformed into HTTP request headers by `GrpcNetClient`.

The last parameter of the `Inject` method is a function that tells the propagator how to inject key-value pairs with trace context into the provided metadata. The propagator, depending on its implementation, may follow different formats and inject different headers. Here, we don't need to worry about it.

Okay, we injected the context to enable correlation with the backend, and now it's time to populate the attributes:

client/GrpcTracingInterceptor.cs

```
private void SetRpcAttributes<Req, Res>(Activity act,
  Method<Req, Res> method)
{
  act.SetTag("rpc.system", "grpc");
  act.SetTag("rpc.service", method.ServiceName);
  act.SetTag("rpc.method", method.Name);
  act.SetTag("net.peer.name", _host);
  if (_port != 80 && _port != 443)
    act.SetTag("net.peer.port", _port);
}
```

https://github.com/PacktPublishing/Modern-Distributed-Trac-ing-in-.NET/blob/main/chapter10/client/GrpcTracingInterceptor.cs

Here, we populate the service and method names from the information provided in the call context. But the host and port come from instance variables we passed to the interceptor constructor – this information is not available in the client interceptor.

Finally, we should populate the gRPC status code and `Activity` status once the response is received:

client/GrpcTracingInterceptor.cs

```
private static void SetStatus(Activity act, Status status)
{
  act.SetTag("rpc.grpc.status_code",
    (int)status.StatusCode);

  var activityStatus = status.StatusCode != StatusCode.OK ?
    ActivityStatusCode.Error : ActivityStatusCode.Unset;

  act.SetStatus(activityStatus, status.Detail);
}
```

https://github.com/PacktPublishing/Modern-Distributed-Trac-
ing-in-.NET/blob/main/chapter10/client/GrpcTracingInterceptor.cs

We left `Activity.Status` unset if the request was successful following the gRPC semantic conventions. It makes sense for generic instrumentation libraries since they don't know what represents a success. In a custom instrumentation, we may know better and can be more specific.

This is it; we just finished unary call instrumentation on the client. Let's now configure a gRPC client to use.

Configuring instrumentation

Let's set up a tracing interceptor on `GrpcClient` instances. In the demo application, we use `GrpcClient` integration with ASP.NET Core and set it up in the following way:

client/Program.cs

```
builder.Services
  .AddGrpcClient<Nofitier.NofitierClient>(o => {
    o => o.Address = serverEndpoint; ... })
  .AddInterceptor(() => new GrpcTracingInterceptor(
    serverEndpoint, contextPropagator))
  ...
```

https://github.com/PacktPublishing/Modern-Distributed-Trac-
ing-in-.NET/blob/main/chapter10/client/Program.cs

Here, we added `GrpcClient`, configured the endpoint, and added a tracing interceptor. We passed the options – the service endpoint and context propagator – explicitly.

The propagator is the implementation of the `TextMapPropagator` class – we use a composite one that supports W3C Trace Context and Baggage formats:

client/Program.cs

```
CompositeTextMapPropagator contextPropagator = new (
  new TextMapPropagator[] {
    new TraceContextPropagator(),
    new BaggagePropagator() });
Sdk.SetDefaultTextMapPropagator(contextPropagator);
```

https://github.com/PacktPublishing/Modern-Distributed-Trac-
ing-in-.NET/blob/main/chapter10/client/Program.cs

The last step is to configure OpenTelemetry and enable `ActivitySource` we use in the interceptor:

client/Program.cs

```
builder.Services.AddOpenTelemetry()
  .WithTracing(b => b.AddSource("Client.Grpc")...);
```

https://github.com/PacktPublishing/Modern-Distributed-Trac-
ing-in-.NET/blob/main/chapter10/client/Program.cs

That's it for the unary client calls. Let's now instrument the server.

Instrumenting server calls

Service-side instrumentation is similar. We can use the gRPC interceptor again and this time override the `UnaryServerHandler` method. Once the request is received, we should extract the context and start a new activity. It should have the `server` kind, a name that follows the same pattern as for the client span – `{package.service}/{method}` – and attributes very similar to those we saw on the client. Here's the interceptor code:

server/GrpcTracingInterceptor.cs

```
var traceContext = _propagator.Extract(default,
  ctx.RequestHeaders,
  static (headers, k) => new[] { headers.GetValue(k) });
Baggage.Current = traceContext.Baggage;
using var activity = Source.StartActivity(ctx.Method,
  ActivityKind.Server, traceContext.ActivityContext);

if (activity?.IsAllDataRequested != true)
  return await continuation(request, ctx);

SetRpcAttributes(activity, ctx.Host, ctx.Method);
try
{
  var response = await continuation(request, ctx);
  SetStatus(activity, ctx.Status);
  return response;
}
catch (Exception ex) {...}
```

https://github.com/PacktPublishing/Modern-Distributed-Trac-
ing-in-.NET/blob/main/chapter10/server/GrpcTracingInterceptor.cs

We extract the trace context and baggage using the propagator and then pass the extracted parent trace context to the new activity and add the attributes. The server interceptor callback is asynchronous, so we can await a response from the server and populate the status.

That's it; now we just need to configure interceptors and enable `ActivitySource`:

server/Program.cs

```
builder.Services
   .AddSingleton<TextMapPropagator>(contextPropagator)
   .AddGrpc(o => {
     o.Interceptors.Add<GrpcTracingInterceptor>(); ...});

builder.Services.AddOpenTelemetry()
     .WithTracing(b => b.AddSource("Server.Grpc")...);
```

https://github.com/PacktPublishing/Modern-Distributed-Trac-
ing-in-.NET/blob/main/chapter10/server/Program.cs

We added gRPC services, configured the tracing interceptor, and enabled the new activity source. It's time to check out generated traces.

Run the application with `$docker-compose up --build` and then hit the frontend at `http://localhost:5051/single/hello`. It will send a message to the server and return a response or show a transient error. An example of a trace with an error is shown in *Figure 10.1*:

Figure 10.1 – gRPC trace showing error on server

Here, we see two spans from the client application and one from the server. They describe an incoming request collected by the ASP.NET Core instrumentation and client and server sides of the gRPC call. *Figure 10.2* shows client span attributes where we can see the destination and status:

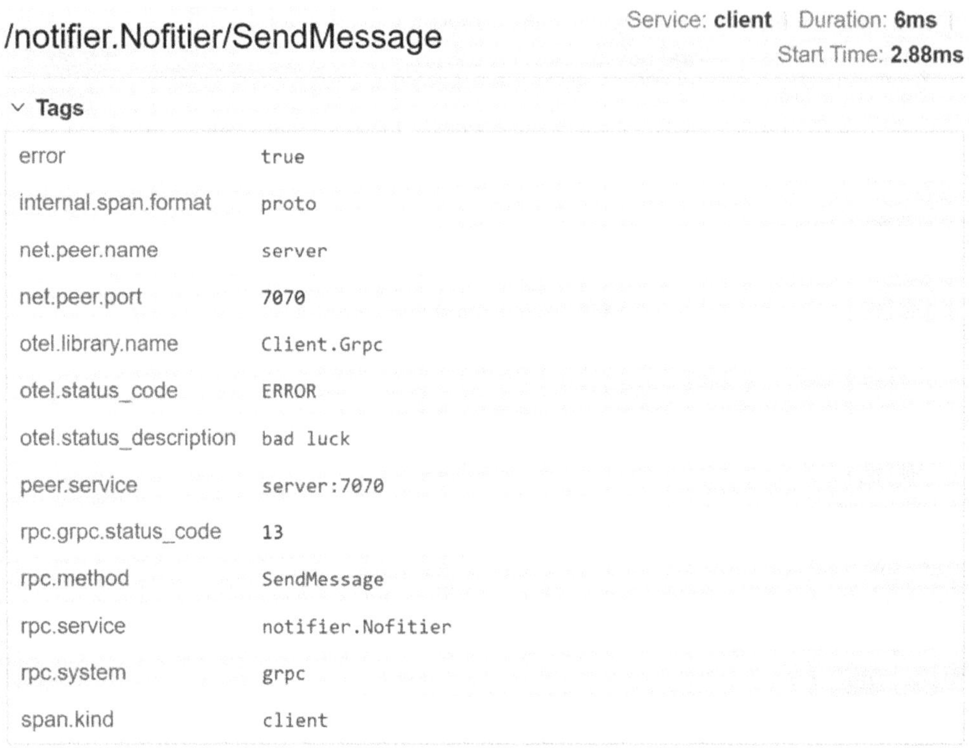

/notifier.Nofitier/SendMessage   Service: **client**   Duration: **6ms**
Start Time: **2.88ms**

˅ **Tags**

error	true
internal.span.format	proto
net.peer.name	server
net.peer.port	7070
otel.library.name	Client.Grpc
otel.status_code	ERROR
otel.status_description	bad luck
peer.service	server:7070
rpc.grpc.status_code	13
rpc.method	SendMessage
rpc.service	notifier.Nofitier
rpc.system	grpc
span.kind	client

Figure 10.2 – gRPC client attributes

This instrumentation allows us to trace unary calls for any gRPC service, which is similar to the HTTP instrumentation we saw in *Chapter 2, Native Monitoring in .NET*. Let's now explore instrumentation for streaming calls.

Instrumenting streaming calls

So far in the book, we have covered the instrumentation of synchronous calls where the application makes a request and awaits its completion. However, it's common to use gRPC or other protocols, such as SignalR or WebSocket, to communicate in an asynchronous way when the client and server establish a connection and then send each other messages.

Common use cases for this kind of communication include chat applications, collaboration tools, and other cases when data should flow in real time and frequently in both directions.

The call starts when the client initiates a connection and may last until the client decides to disconnect, the connection becomes idle, or some network issue happens. In practice, it means that such calls may last for days.

While a connection is alive, the client and server can write each other messages to corresponding network streams. It's much faster and more efficient when the client and server communicate frequently within a relatively short period of time. This approach minimizes the overhead created by DNS lookup, protocol negotiation, load balancing, authorization, and routing compared to request-response communication when at least some of these operations would happen for each request.

On the downside, the application could become more complex as in many cases, we'd still need to correlate client messages with service replies to them and come up with our own semantics for metadata and status codes.

For observability, it means that out-of-the-box instrumentation is rarely enough and at least some custom instrumentation is necessary. Let's see why.

Basic instrumentation

Some applications pass completely independent messages within one streaming call and would want different traces to describe individual messages. Others use streaming to send scoped batches of messages and would rather expect one trace to describe everything that happens within one streaming call. When it comes to streaming, there is no single solution.

gRPC auto-instrumentations follow OpenTelemetry semantic conventions and provide a default experience where a streaming call is represented with client and server spans, even if the call lifetime is unbound. Individual messages are described with span events with attributes covering the direction, message identifier, and size.

You can find a full instrumentation implementation that follows these conventions in the `client/GrpcTracingInterceptor.cs` and `server/GrpcTracingInterceptor.cs` files in the book's repository. Let's look at the traces it produces.

Go ahead and start the application with `$ docker-compose up --build` and then hit the client application at `http://localhost:5051/streaming/hello?count=2`. It will send two messages to the server and read all the responses.

Check out Jaeger at `http://localhost:16686/`. You should see a trace similar to the one shown in *Figure 10.3*:

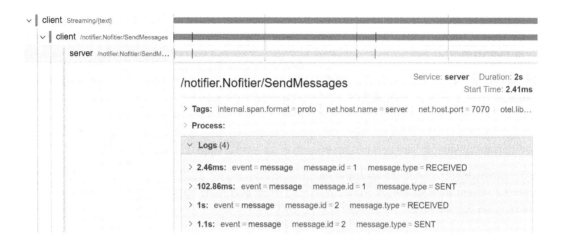

Figure 10.3 – Streaming call with events

Similarly to a unary call, the trace consists of three spans. The only difference is that client and server gRPC spans have events – two events per message, indicating when the message was sent and received. The message.id attribute here represents the sequence number of a message in a request or response stream and might be used to correlate request and response messages.

The trace shown in *Figure 10.3* represents the best we can achieve with auto-instrumentation that is not aware of our specific streaming usage. Let's see how we can improve it.

Tracing individual messages

Let's pretend that the client initiates a very long stream – in this case, the previous trace would not be very helpful. Assuming messages are not too frequent and verbose, we might want to instrument each specific message and see how server response messages correlate with client messages.

To instrument individual messages, we'd have to propagate context inside the message, which is not possible in an interceptor where we operate with generic message types.

Our message protobuf definition contains text and an attribute map that we can use to pass trace context:

client\Protos\notifier.proto

```
message Message {
    string text = 1;
    map<string, string> attributes = 2;
}
```

https://github.com/PacktPublishing/Modern-Distributed-Tracing-in-.NET/blob/main/chapter10/client/notifier.proto

We're going to create one client span per message to describe and identify it, and a server span that will represent processing the message.

If we have hundreds of messages during one streaming call, having all of them in one trace will be hard to read. Also, typical sampling techniques would not apply – depending on the sampling decision made for the whole streaming call, we'll sample in or drop all per-message spans.

Ideally, we want to have a trace per message flow and have a link to the long-running HTTP requests that carried the message over. This way, we still know what happened with the transport and what else was sent over the same HTTP request, but we'll make independent sampling decisions and will have smaller and more readable traces.

> **Note**
>
> Tracing individual messages is reasonable when messages are relatively big and processing them takes a reasonable amount of time. Alternative approaches may include custom correlation or context propagation for sampled-in messages only.

Let's go ahead and instrument individual messages: we'll need to start a new activity per message with the `producer` kind indicating an async call. We need to start a new trace and use `Activity.Current` as a link rather than a parent:

client/controllers/StreamingController.cs

```
IEnumerable<ActivityLink>? links = null;
if (Activity.Current != null)
{
  links = new[] {
    new ActivityLink(Activity.Current.Context) };
  Activity.Current = null;
}
using var act = Source.StartActivity("SendMessage",
  ActivityKind.Producer,
  default(ActivityContext),
  links: links);
```

https://github.com/PacktPublishing/Modern-Distributed-Tracing-in-.NET/blob/main/chapter10/client/Controllers/StreamingController.cs

We created a link from the current activity and then set `Activity.Current` to `null`, which forces the `StartActivity` method to create an orphaned activity.

> **Note**
>
> Setting `Activity.Current` should be done with caution. In this example, we're starting a new task specifically to ensure that it won't change the `Activity.Current` value beyond the scope of this task.

We have an activity; now it's time to inject the context and send a message to the server:

client/controllers/StreamingController.cs

```
_propagator.Inject(
  new PropagationContext(act.Context, Baggage.Current),
  message,
  static (m, k, v) => m.Attributes.Add(k, v));
try
{
  await requestStream.WriteAsync(message);
}
catch (Exception ex)
{
  act?.SetStatus(ActivityStatusCode.Error, ex.Message);
}
```

https://github.com/PacktPublishing/Modern-Distributed-Trac-ing-in-.NET/blob/main/chapter10/client/Controllers/StreamingCon-troller.cs

Context injection looks similar to what we did in the client interceptor earlier in this chapter, except here we inject it into message attributes rather than gRPC call metadata.

On the server side, we need to extract the context from the message, then use it as a parent. We should also set `Activity.Current` as a link so we don't lose correlation between the message processing and streaming calls. The new activity has a `consumer` kind, which indicates the processing side of the async call, as shown in this code snippet:

server/NotifierService.cs

```
var context = _propagator.Extract(default,
  message,
  static (m, k) => m.Attributes.TryGetValue(k, out var v)
    ? new [] { v } : Enumerable.Empty<string>());
var link = Activity.Current == null ?
  default : new ActivityLink(Activity.Current.Context);
```

```
using var act = Source.StartActivity(
  "ProcessMessage",
  ActivityKind.Consumer,
  context.ActivityContext,
  links: new[] { link });
...
```

https://github.com/PacktPublishing/Modern-Distributed-Trac-ing-in-.NET/blob/main/chapter10/server/NotifierService.cs

We can now enable corresponding client and server activity sources – we used different names for per-message tracing and interceptors, so we can now control instrumentations individually. Go ahead and enable `Client.Grpc.Message` and `Server.Grpc.Message` sources on the client and server correspondingly and then start an application.

If we hit the streaming endpoint at `http://localhost:5051/streaming/hello?count=5` and then went to Jaeger, we'd see six traces – one for each message sent and one for the gRPC call.

Per-message traces consist of two spans, like the one shown in *Figure 10.4*:

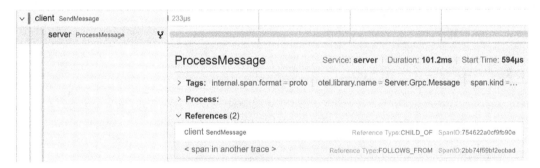

Figure 10.4 – Tracing messages in individual traces

Here, we see that sending this message took about 1 ms and processing it took about 100 ms. Both spans have links (references in Jaeger terminology) to spans describing the client and server sides of the underlying gRPC call.

If we didn't force new trace creation for individual messages, we'd see only one trace containing all the spans, as shown in *Figure 10.5*:

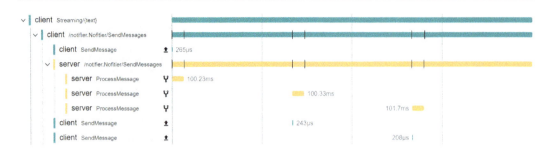

Figure 10.5 – Tracing a streaming call with all messages in one trace

Depending on your scenario, you might prefer to separate traces, have one big trace, or come up with something else.

Note that we can now remove our custom tracing interceptor and enable shared gRPC and HTTP client instrumentation libraries. If you do this, the per-message instrumentation will remain exactly the same and will keep working along with auto-instrumentation.

With this, you should be able to instrument unary or streaming gRPC calls and have an idea of how to extend it to other cases, including SignalR or socket communication.

Let's now see how to use gRPC instrumentation to investigate issues.

Observability in action

There are several issues in our server application. First issue reproduces sporadically when you hit the frontend at `http://localhost:5051/single/hello` several times. You might notice that some requests take longer than others. If we look at the duration metrics or Jaeger's duration view, we'll see something similar to *Figure 10.6*:

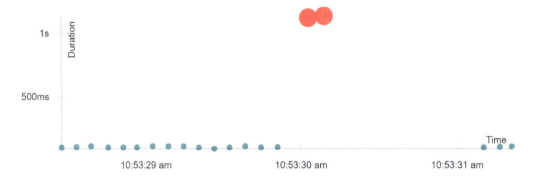

Figure 10.6 – Duration view in Jaeger

We see that most of the calls are fast (around 100 ms), but there is one that takes longer than a second. If we click on it, Jaeger will open the corresponding trace, like the one shown in *Figure 10.7*:

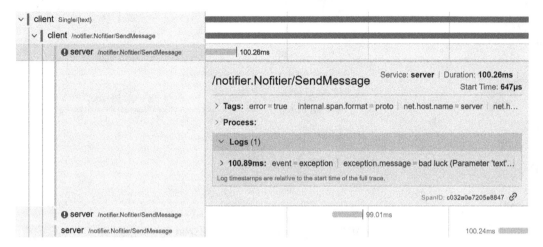

Figure 10.7 – Long trace with errors

Apparently, there were three attempts to send the message – the first two were not successful, but the third one succeeded. So retries are the source of long latency. We can investigate the error by expanding the exception event – we'll see a full stack trace there.

Notably, we see retries only on the service side here. There is just one gRPC span on the client side. What happens here is that we enable a retry policy on the gRPC client channel, which internally adds a retry handler to the `HttpClient` level. So, our tracing interceptor is not called on tries and traces the logical part of the gRPC call.

The official `OpenTelemetry.Instrumentation.GrpcNetClient` instrumentation works properly and traces individual tries on the client as well.

Let's look at another problem. Send the following request: `http://localhost:5051/streaming/hello?count=10`. It will return a few messages and then stop. If we look into Jaeger traces, we'll see a lot of errors for individual messages. Some of them will have just a client span, like the one shown in *Figure 10.8*:

Figure 10.8 – Client error without server span

There is not much information in the span, but luckily, we have a link to the gRPC call. Let's follow it to see whether it explains something. The corresponding trace is shown in *Figure 10.9*:

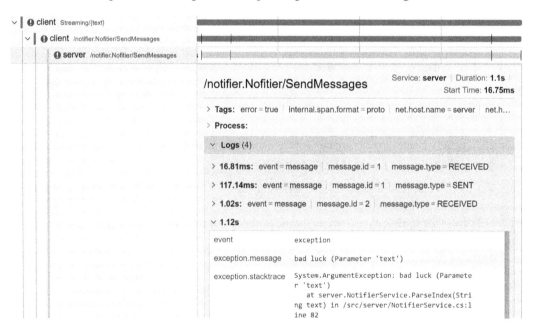

Figure 10.9 – Error in the middle of the gRPC stream

Here, we see a familiar trace, but the processing has failed while parsing the message text. The server span has six events, indicating that two messages were received and the response was successfully sent to the server. The third one was received but then instead of the response, we see an exception with a stack trace to help us investigate further.

If we expand the client gRPC span, we'll see more exceptions for each message that was attempted to be sent after the server error has happened.

But there were no retries – why? In our case, gRPC retries, as we've seen in the previous example, are applied on the HTTP level. In the case of streaming, it means that after the first response is received from the server, the HTTP response, including status codes and headers, is received and the rest of the communication happens within the request and response streams. You can read more about this in the Microsoft gRPC documentation at `https://learn.microsoft.com/en-us/aspnet/core/grpc/retries`.

So, once an unhandled exception is thrown on the server for a particular message, it ends the gRPC call and corresponding request and response streams on the client and server. It affects all remaining messages on the client and explains the partial response we noticed.

Distributed tracing helps us see what happens and learn more about the technologies we use. In addition to tracing, OpenTelemetry defines a set of metrics to monitor on the client and server sides, which includes the duration, the failure rate that can be derived from it, the number of requests and responses, and the payload sizes.

Summary

In this chapter, we got hands-on experience in instrumenting network calls using gRPC as an example. Before starting instrumentation, we learned about the available instrumentation libraries and what OpenTelemetry semantic conventions recommend recording for gRPC.

First, we instrumented unary calls with client and server spans and propagated context through gRPC metadata. Then, we experimented with gRPC streaming, which needs a different approach to tracing. The generic instrumentation of streaming calls suggests creating an event per individual request and response message in the stream and provides a basic level of observability. Depending on our scenarios and observability needs, we can add another layer of instrumentation to trace individual messages. These custom spans work on top of the generic gRPC instrumentation.

Finally, we used tracing to get insights into high latency and transient error scenarios, which also helped us understand gRPC internals.

You're now ready to instrument your network stack with tracing or enrich existing instrumentation libraries by adding custom layers of instrumentation specific to your application. In the next chapter, we'll look into messaging scenarios and dive even deeper into observability for asynchronous processing.

Questions

1.   When using gRPC, would you write your own instrumentation or reuse an existing one?

2.   Let's imagine we want to instrument gRPC communication between the client and server when the client initiates a connection at startup time and keeps it open forever (until the server or client stops) and then reuses this connection for all the communication. Which tracing approach would you choose? Why?

11

Instrumenting Messaging Scenarios

Messaging and asynchronous processing improve distributed system scalability and reliability by reducing coupling between services. However, they also increase complexity and introduce a new failure mode, which makes observability even more important.

In this chapter, we'll work on instrumenting a messaging producer and consumer with traces and metrics and cover individual and batch message processing.

In this chapter, you'll learn how to do the following:

- Trace individual messages as they are created and published
- Instrument receiving and processing operations
- Instrument batches
- Use instrumentation to diagnose common messaging problems

By the end of this chapter, you should be able to instrument your messaging application from scratch or tune the existing messaging instrumentation to your needs.

Technical requirements

The code for this chapter is available in the book's repository on GitHub at `https://github.com/PacktPublishing/Modern-Distributed-Tracing-in-.NET/tree/main/chapter11`.

To run the samples and perform analysis, we'll need the following tools:

- .NET SDK 7.0 or later.

- Docker and `docker-compose`.

- Any HTTP benchmarking tool, for example, `bombardier`. You can install it with `$ go get -u github.com/codesenberg/bombardier` if you have Go tools, or download bits directly from its GitHub repository at `https://github.com/codesenberg/bombardier/releases`.

We will also be using the Azure Storage emulator in Docker. No setup or Azure subscription is necessary.

Observability in messaging scenarios

In *Chapter 10*, *Tracing Network Calls*, we just started scratching the surface of tracing support for asynchronous processing. There, we saw how the client and server can send a stream of potentially independent messages to each other.

In the case of messaging, things get even more complicated: in addition to asynchronous communication, the producer and consumer interact through an intermediary – a messaging **broker**.

Operation on the producer completes once the message is published to the broker without waiting for the consumer to process this message. Depending on the scenario and application health, the consumer may process it right away, in a few seconds, or in several days.

In some cases, producers get a notification that the message was processed, but this usually happens through another messaging queue or a different communication channel.

Essentially, the producer does not know whether the consumer exists – failures or delays in the processing pipeline are not visible on the producer side. This changes how we should look at latency, throughput, or error rate from an observability standpoint – now we need to think about end-to-end flows that consist of multiple independent operations.

For example, when using HTTP calls only, the latency of the original request covers almost everything that happened with the request. Once we introduce messaging, we need means to measure the end-to-end latency and identify failures between different components. An example of an application that uses messaging is shown in *Figure 11.1*:

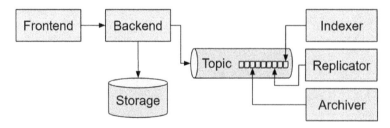

Figure 11.1 – Application using messaging to run tasks in the background

In such an application, when the user sends a request to the frontend, they receive a response once the backend finishes processing and publishes a message to a topic. The indexer, replicator, archiver, and any other services that post-process the data run at their own speed. The indexer usually processes the latest messages, while the archiver would only look at the messages published days ago.

Some of these components can fail without affecting user scenarios directly, while others impact how soon the data the user published shows up in other parts of the system and therefore can be critical.

Let's explore how we can instrument such applications.

Before writing our own instrumentation from scratch, we should always check whether there are existing instrumentation libraries we can start with, and if there are none available, we should consult with OpenTelemetry semantic conventions.

We're going to instrument Azure Queue Storage as an example. The existing instrumentation does not cover the messaging aspects of queues because of the reasons we'll see in the next couple of sections. So, we'll have to write our own; we'll do it according to OpenTelemetry conventions.

Messaging semantic conventions

The messaging conventions for tracing are available at `https://github.com/open-telemetry/opentelemetry-specification/blob/main/specification/trace/semantic_conventions/messaging.md`.

They currently have experimental status and are very likely to change. There are no general metrics conventions available yet, but you can find ones specific to Kafka.

Conventions provide some considerations on context propagation (we'll discuss this in the *Trace context propagation* section) and define generic attributes to describe messaging operations. Here are a few essential ones we're going to use:

- `messaging.system`: Indicates that the span follows messaging semantics and describes the specific messaging system used, such as `kafka` or `rabbitmq`. In our sample, we'll use `azqueues`.
- `messaging.operation`: Identifies one of the standard operations: `publish`, `receive`, or `process`.
- `messaging.destination.name` and `messaging.source.name`: Describe a queue or topic name within a broker. The term `destination` is used on the producer and `source` is used on the consumer.
- `net.peer.name`: Identifies the broker domain name.

Let's see how we can use the conventions to add observability signals that can help us document the application behavior or detect and resolve a new class of issues happening in messaging scenarios.

Instrumenting the producer

The producer is the component responsible for publishing messages to a broker. The publishing process itself is usually synchronous: we send a request to the broker and get a response from it indicating whether the message was published successfully.

Depending on the messaging system and producer needs, one publish request may carry one or more messages. We'll discuss batching in the *Instrumenting batching scenarios* section. For now, let's focus on a single message case.

To trace it, we need to make sure we create an activity when we publish a message, so we can track the call duration and status and debug individual requests. We'd also be interested in metrics for duration, throughput, and failure rate – it's important to budget cloud messaging solutions or scale self-hosted brokers.

Another essential part of producer instrumentation is context propagation. Let's stop here for a second and discuss it.

Trace context propagation

When we instrument HTTP calls, context is propagated via HTTP request headers, which are part of the request. In messaging, the context is carried via a transport call to the broker and is not propagated to a consumer. Transport call trace context identifies the request, but not the message(s) it carries.

So, we need to propagate context inside the message to make sure it goes all the way to the consumer. But which context should we inject? We have several options:

- **Use context from the current activity**: For instance, when we publish messages in the scope of an incoming HTTP request, we may use the context of the activity representing this HTTP server call. This works only if we send one message per incoming request. If we send more than one (each in an individual publish call), we'd not be able to tell which message the consumer call processed or identify whether we sent messages to the right queues.

- **Create an activity per message and inject its context**: Unique context allows us to trace messages individually and works in batching scenarios as well where we send multiple messages in one publish call. It also adds the overhead of creating an additional activity per message.

- **Reuse the publish activity**: When we publish one message in one call to the broker, we can uniquely identify a message and publish call with one activity.

The first option goes against OpenTelemetry messaging semantic conventions, which allow us to pick a suitable option from the last two. In our example, we're using Azure Queue Storage, which does not support batching when publishing messages. So, we're going to use the last option and create one activity to trace a publish call and inject its context into the message.

> **Note**
>
> When forking or routing messages from one queue to another, the message might have pre-existing trace context injected in the upstream service. The default behavior in such a case should be to keep the message context intact. To correlate all operations that happen with the message, we can always add a link to an existing trace context in the message when publishing or receiving it.

Another interesting aspect of Azure Queue Storage is that it doesn't support message metadata – the message is an opaque payload without any prescribed structure or format that the service carries over. So, similarly to gRPC streaming, which we covered in *Chapter 10, Tracing Network Calls*, we'll need to define our own message structure or use one of the well-known event formats available out there, such as **CloudEvents**.

> **Note**
>
> CloudEvents (`https://cloudevents.io`) is an open standard that defines event structure in a vendor- and technology-agnostic way. It's commonly used by cloud providers to notify applications about infrastructure changes or when implementing data change feeds. CloudEvents have distributed tracing extensions to carry W3C Trace Context as well as general-purpose metadata that can be used for other formats. OpenTelemetry also provides semantic conventions for CloudEvents.

For demo purposes, we'll keep things simple and define our own tiny message model in the following way:

producer/Message.cs

```
public class Message
{
    ...
    public Dictionary<string, string> Headers { get; set; } =
        new ();
    public string? Text { get; set; }
}
```

https://github.com/PacktPublishing/Modern-Distributed-Tracing-in-.NET/blob/main/chapter11/producer/Message.cs

We'll use the `Headers` property to propagate the trace context and will keep the payload in the `Text` property.

Similarly to the gRPC streaming examples we saw in *Chapter 10, Tracing Network Calls*, we can inject context into this message using the OpenTelemetry propagator with the following code snippet:

producer/Controllers/SendController.cs

```
private void InjectContext(Message message, Activity? act)
{
  if (act != null)
  {
    _propagator.Inject(new (act.Context, Baggage.Current),
      message,
      static (m, k, v) => m.Headers[k] = v);
  }
}
```

https://github.com/PacktPublishing/Modern-Distributed-Tracing-in-.NET/blob/main/chapter11/producer/Controllers/SendController.cs

Now we have all we need to instrument a publish call – let's do it.

Tracing a publish call

We'll need to create a new activity and put common messaging attributes on it to identify the broker, queue operation, and add other information. In the case of Azure Queue Storage, we can use the account name as the broker identifier (as they are unique within a public cloud).

Then, we'll inject context into the message and proceed with publishing. After the message is published successfully, we can also record the information returned by the broker, such as the message ID and other details we might consider useful.

Here's the corresponding code:

producer/Controllers/SendController.cs

```
Stopwatch? duration = PublishDuration.Enabled ?
  Stopwatch.StartNew() : null;

using var act = StartPublishActivity();
InjectContext(message, Activity.Current);
try
{
  var receipt = await _queue.SendMessageAsync(
    BinaryData.FromObjectAsJson(message));
```

```
  act?.SetTag("messaging.message.id",
    receipt.Value.MessageId);
  RecordPublishMetrics(duration, "ok");

  ...
}
catch (Exception ex)
{
  act?.SetStatus(ActivityStatusCode.Error, ex.Message);
  RecordPublishMetrics(duration, "fail")

  ...
}
```

https://github.com/PacktPublishing/Modern-Distributed-Trac-
ing-in-.NET/blob/main/chapter11/producer/Controllers/SendControl-
ler.cs

Here, we injected the context of `Activity.Current` with the `Inject` method we implemented before. This may be useful if you want to turn off per-message activities. In such a case, per-message tracing will be limited, but consumer and producer calls will still be correlated. We also record metrics here – stay tuned for the details; we're going to cover them in the next section.

Here's the `StartPublishActivity` method implementation:

producer/Controllers/SendController.cs

```
var act = Source.StartActivity($"{_queue.Name} publish",
  ActivityKind.Producer);
if (act?.IsAllDataRequested == true)
  act.SetTag("messaging.system", "azqueues")
    .SetTag("messaging.operation", "publish")
    .SetTag("messaging.destination.name", _queue.Name)
    .SetTag("net.peer.name", _queue.AccountName)
}
```

https://github.com/PacktPublishing/Modern-Distributed-Trac-
ing-in-.NET/blob/main/chapter11/producer/Controllers/SendControl-
ler.cs

The activity here has a `producer` kind, which indicates the start of an async flow. The name follows OpenTelemetry semantic conventions, which recommend using the `{queue_name} {operation}` pattern. We can also cache it to avoid unnecessary string formatting.

This is it; we've covered producer tracing – let's look at metrics now.

Producer metrics

Messaging-specific metrics come as an addition to resource utilization, .NET runtime, HTTP, and other metrics you might want to expose.

To some extent, we can use HTTP metrics to monitor calls to Azure Queue Storage since they work on top of HTTP. This would allow us to monitor duration, success rate, and throughput for individual HTTP calls to storage, but won't allow us to distinguish queues within one storage account.

So, if we rely on metrics, we should record some messaging-specific ones that cover common indicators such as publish call duration, throughput, and latency for each queue we use.

We can report all of them using a duration histogram, as we saw in *Chapter 7, Adding Custom Metrics*. First, let's initialize the duration histogram, as shown in the following code snippet:

producer/Controllers/SendController.cs

```
private static readonly Meter Meter = new("Queue.Publish");
private static readonly Histogram<double> PublishDuration =
  Meter.CreateHistogram<double>(
    "messaging.azqueues.publish.duration", ...);
```

https://github.com/PacktPublishing/Modern-Distributed-Tracing-in-.NET/blob/main/chapter11/producer/Controllers/SendController.cs

`Meter` and `Histogram` are static since we defined them in the controller. The controller lifetime is scoped to a request, so we keep them static to stay efficient.

As we saw in the tracing example, every time we publish a message, we're also going to record a publish duration. Here's how it's implemented:

producer/Controllers/SendController.cs

```
public void RecordPublishMetrics(Stopwatch? dur,
  string status)
{
  ...
  TagList tags = new() {
    { "net.peer.name", _queue.AccountName },
    { "messaging.destination.name", _queue.Name },
    { "messaging.azqueue.status", status }};

  PublishDuration.Record(dur.Elapsed. TotalMilliseconds,
    tags);
}
```

```
https://github.com/PacktPublishing/Modern-Distributed-Trac-
ing-in-.NET/blob/main/chapter11/producer/Controllers/SendControl-
ler.cs
```

Here, we used the same attributes to describe the queue and added a custom status attribute. Keep in mind that we need it to have low cardinality, so we only use `ok` and `fail` statuses when we call this method.

We're done with the producer. Having basic tracing and metrics should give us a good starting point to diagnose and debug most of the issues and monitor overall producer health, as we'll see in the *Performance analysis in messaging scenarios* section later. Let's now explore instrumentation on consumers.

Instrumenting the consumer

While you might be able to get away without custom instrumentation on the producer, consumer instrumentation is unavoidable.

Some brokers push messages to consumers using synchronous HTTP or RPC calls, and the existing framework instrumentation can provide the bare minimum of observability data. In all other cases, messaging traces and metrics are all we have to detect consumer health and debug issues.

Let's start by tracing individual messages – recording when they arrive in the consumer and how they are processed. This allows us to debug issues by answering questions such as "Where is this message now?" or "Why did it take so long to process the data?"

Tracing consumer operations

When using Azure Queue Storage, applications request one or more messages from the queue. Received messages stay in the queue but become invisible to other consumers for configurable visibility timeout. The application processes messages and, when done, deletes them from the queue. If processing fails with a transient issue, applications don't delete messages. The same flow is commonly used when working with AWS SQS.

RabbitMQ- and AMQP-based messaging flows look similar, except messages can be pushed to the consumer so that the application reacts to the client library callback instead of polling the queue.

Callback-based delivery allows us to implement instrumentation in client libraries or provide a shared instrumentation library, and with a poll-based model, we essentially are forced to write at least some custom instrumentation for processing. Let's do it.

First, let's instrument message processing in isolation from receiving. We'll need to create an activity to track processing that will capture everything that happens there, including message deletion:

consumer/SingleReceiver.cs

```
using var act = StartProcessActivity(msg);
...
try
{
  await ProcessMessage(msg, token);
  await _queue.DeleteMessageAsync(msg.MessageId,
    msg.PopReceipt, token);
}
catch (Exception ex)
{
  await _queue.UpdateMessageAsync(msg.MessageId,
    msg.PopReceipt, visibilityTimeout: BackoffTimeout,
    cancellationToken: token);
  ...
  act?.SetStatus(ActivityStatusCode.Error, ex.Message);
}
```

https://github.com/PacktPublishing/Modern-Distributed-Trac-ing-in-.NET/blob/main/chapter11/consumer/SingleReceiver.cs

Here, all the processing logic happens in the `ProcessMessage` method. When it completes successfully, we delete the message from the queue. Otherwise, we update its visibility to reappear in the queue after the backoff timeout.

Here's the `StartProcessActivity` implementation:

consumer/SingleReceiver.cs

```
PropagationContext ctx = ExtractContext(msg);
var current = new ActivityLink(Activity.Current?.Context ??
  default);
var act = _messageSource.StartActivity(
  $"{_queue.Name} process",
  ActivityKind.Consumer,
  ctx.ActivityContext,
  links: new[] { current });
if (act?.IsAllDataRequested == true)
  act.SetTag("net.peer.name",_queue.AccountName)
```

```
    .SetTag("messaging.system", "azqueues")
    .SetTag("messaging.operation", "process")
    .SetTag("messaging.source.name", _queue.Name)
    .SetTag("messaging.message.id", msg.MessageId);
...
```

https://github.com/PacktPublishing/Modern-Distributed-Trac-
ing-in-.NET/blob/main/chapter11/consumer/SingleReceiver.cs

Here, we extracted the context from the message and used it as a parent of the processing activity. It has the `consumer` kind, which indicates the continuation of the asynchronous flow. We also kept `Activity.Current` as a link to preserve correlation. We also added messaging attributes.

Message deletion and updates are traces by HTTP or Azure Queue SDK instrumentations. They don't have messaging semantics, but should give us reasonable observability. Corresponding activities become children of the processing one, as shown in *Figure 11.2*:

Figure 11.2 – Message trace from producer to consumer

The message was published and then we see two attempts to process it on the consumer: the first attempt failed. The second try was successful, and the message was deleted.

What's missing in the preceding screenshot? We don't see how and when the message was received. This might not be important on this trace, but look at another one in *Figure 11.3*:

Figure 11.3 – Message trace with a nine-minute gap between producer and consumer

Here, nothing has happened for almost nine minutes. Was the message received by a consumer during that time? Were the consumers alive? What were they doing? Were there any problems in the Azure Queue service that prevented messages from being received?

We'll see how to answer these questions later. Now, let's focus on tracing the receive operation.

The challenge with the receive operation is that the message trace context is available after the message is received and the corresponding operation is about to end. We could add links to message trace contexts then, but it's currently only possible to add them at activity start time.

This is likely to change, but for now, we'll work around it by tracing the receive-and-process iteration and adding an attribute with the received message ID so we can find all spans that touched this message:

consumer/SingleReceiver.cs

```
using var act = _receiverSource
  .StartActivity("ReceiveAndProcess");
try
{
  var response = await _queue.ReceiveMessagesAsync(1,
    ProcessingTimeout, token);
  QueueMessage[] messages = response.Value;
  if (messages.Length == 0)
  {
    ...; continue;
  }
  act?.SetTag("messaging.message.id",
    messages[0].MessageId);
  await ProcessAndSettle(messages[0], token);
  ...
}
catch (Exception ex)
{
  act?.SetStatus(ActivityStatusCode.Error, ex.Message);
  ...
}
```

https://github.com/PacktPublishing/Modern-Distributed-Trac-ing-in-.NET/blob/main/chapter11/consumer/SingleReceiver.cs

Here, we receive at most one message from the queue. If a message was received, we process it.

One iteration is tracked with the ReceiveAndProcess activity, which becomes a parent to the receiving operation. The message processing activity is created in the ProcessAndSettle method and links to the ReceiveAndProcess activity, as shown in *Figure 11.4*:

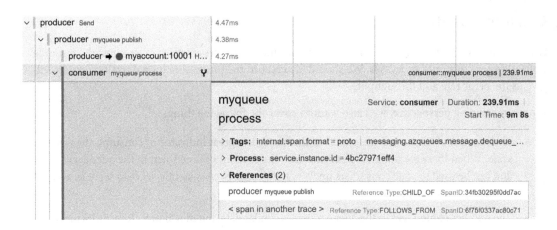

Figure 11.4 – Link from processing to outer loop activity

If we follow the link, we'll see an outer loop trace like the one shown in *Figure 11.5*:

Figure 11.5 – Trace representing the receive and process iteration

Since more and more observability backends are providing better support for links, it can be more convenient to use them in your backend.

With iteration instrumented, we can now correlate receiving and processing or see how long a full loop cycle takes. This can help us understand whether consumers are alive and trying to receive and process something.

We're stamping the `messaging.message.id` attribute on all spans to simplify finding all operations related to any given message.

Now, back to the nine-minute gap we saw in *Figure 11.3*. What happened there is that we got too many messages in the queue – they were produced faster than we consumed them. By looking at gaps in individual traces, we can suspect that message spent time in the queue, but can't tell for sure. What we need is to see the rate at which messages are published, processed, and deleted. We should also understand how long messages spend in the queue and how big the queue is. Let's see how we can record and use such metrics.

Consumer metrics

Similar to producers, we should enable common runtime and process metrics so we know the resource utilization for consumer processes. We should also record the processing loop duration, which will give us the error rate and throughput.

From a messaging perspective, we'd also want to cover the following things:

- The amount of time messages spend in the queue is a great indicator of consumer health and scale. When there are not enough consumers, the amount of time spent in the queue will grow and can be used to scale consumers up. When it decreases consistently, it could serve as a signal to scale consumers down.

- The number of messages in the queue provides similar data, but in real time. It includes messages that have not yet been processed. Queue size metric can also be recorded on the producer side without ever depending on the consumer.

These metrics, or similar ones you can come up with, and their trends over time provide a great indication of consumer health.

These metrics increase if consumer performance degrades or the error rate increases. They won't be helpful if consumers fail to process messages but immediately delete them from the queue, but this will manifest in high error rate. So, let's go ahead and instrument our application with these metrics.

Duration, throughput, and failure rate

We're going to measure the processing loop duration, which includes trying to receive a message and its processing. Measuring the receiving and processing duration independently would be even more precise and is something to consider in your production applications.

At the beginning of the loop, we'll start a stopwatch to measure operation duration, and once processing completes, we'll report it as a histogram along with queue information and the status. Let's first create the histogram instrument:

---

consumer/SingleReceiver.cs

```
private readonly Meter _meter = new ("Queue.Receive");
private readonly Histogram<double> _loopDuration;
...
_loopDuration = _meter.CreateHistogram<double>(
  "messaging.azqueues.process.loop.duration", "ms",
  "Receive and processing duration.");
```

https://github.com/PacktPublishing/Modern-Distributed-Trac-ing-in-.NET/blob/main/chapter11/consumer/SingleReceiver.cs

---

We create meter and duration instruments here as instance variables, which we dispose of along with the `SingleReceiver` instance. The receiver extends the `BackgroundService` interface and is registered in the dependency injection container as a singleton, so they are all disposed of once the application shuts down.

The processing loop instrumentation can be done in the following way:

consumer/SingleReceiver.cs

```
Stopwatch? duration = Stopwatch.StartNew();
try
{
  var response = await _queue.ReceiveMessagesAsync(1,
    ProcessingTimeout, token);
  QueueMessage[] messages = response.Value;
  RecordLag(messages);
  if (messages.Length == 0)
  {

    ...
    RecordLoopDuration(duration, "empty");
    continue;
  }
  ...
  await ProcessAndSettle(messages[0], token);
  RecordLoopDuration(duration, "ok");
}
catch (Exception ex)
{
  RecordLoopDuration(duration, "fail"); ...
}
```

https://github.com/PacktPublishing/Modern-Distributed-Trac-ing-in-.NET/blob/main/chapter11/consumer/SingleReceiver.cs

Here, we record the duration of each iteration along with the queue information and status. The status can have the following values: `ok`, `fail`, or `empty` (if no messages were received). In real applications, you probably want to be more precise and add a few more statuses to indicate the failure reason. For example, it would be important to record why the receive operation failed, whether there was a serialization or validation error, processing timed out, or it failed with a terminal or transient error.

The `RecordLoopDuration` method implementation is shown in this snippet:

---

consumer/SingleReceiver.cs

```
TagList tags = new () {
  { "net.peer.name", _queue.AccountName },
  { "messaging.source.name", _queue.Name },
  { "messaging.azqueue.status", status }};


_loopDuration.Record(duration.ElapsedMilliseconds, tags);
```

https://github.com/PacktPublishing/Modern-Distributed-Trac-ing-in-.NET/blob/main/chapter11/consumer/SingleReceiver.cs

---

We'll see how we can use this metric later in this chapter. Let's first implement consumer lag and queue size.

Consumer lag

In the code sample showing metrics in the processing loop, we called into the `RecordLag` method as soon as we received a message. Consumer lag records the approximate time a message spent in the queue – the delta between the receive and enqueue time.

The enqueue time is recorded by the Azure Queue service and is exposed as a property on the `QueueMessage` instance. We can record the metric with the following code:

---

consumer/SingleReceiver.cs

```
_consumerLag = _meter.CreateHistogram<double>(
  "messaging.azqueues.consumer.lag", "s", ...);
...
long receivedAt = DateTimeOffset.UtcNow
  .ToUnixTimeMilliseconds();

TagList tags = new () {
  { "net.peer.name", _queue.AccountName },
  { "messaging.source.name", _queue.Name }};

foreach (var msg in messages
    .Where(m => m.InsertedOn.HasValue))
{
  long insertedOn = msg.InsertedOn!
    .Value.ToUnixTimeMilliseconds());
```

```
    long lag = Math.Max(1, receivedAt - insertedOn);
    _consumerLag.Record(lag/1000d, tags);
}
```

https://github.com/PacktPublishing/Modern-Distributed-Trac-ing-in-.NET/blob/main/chapter11/consumer/SingleReceiver.cs

Here, we create a histogram that represents the lag (in seconds) and record it for every received message as the difference between the current time and the time at which the message was received by the broker.

Note that these timestamps usually come from two different computers – the difference can be negative and is not precise due to clock skew. The margin of error can reach seconds but may, to some extent, be corrected within your system.

Clock skew should be expected, but sometimes things can go really wrong. I once was involved in investigating an incident that took our service down in one of the data centers. It happened because of the wrong time server configuration, which moved the clock on one of the services back a few hours. It broke authentication – the authentication tokens had timestamps from hours ago and were considered expired.

Despite being imprecise, consumer lag should give us an idea of how long messages spend in a queue. We record it every time a message is received, so it also reflects redeliveries. Also, we record it before we know whether processing was successful, so it does not have any status.

Before we record lag on the consumer, we first need to receive a message. When we see a huge lag, it's a good signal that something is not right, but it does not tell us how many messages have not yet been received.

For example, when the load is low and the queue is empty, there might be a few invalid messages that are stuck there. It's a bug, but it can be fixed during business hours. To detect how big the issue is, we also need to know how many messages are in the queue. Let's see how to implement it.

Queue size

Azure Queue Storage as well as Amazon SQS allow us to retrieve an approximate count of messages. We can register another `BackgroundService` implementation to retrieve the count periodically. This can be done on the consumer or producer.

We'll use a gauge instrument to report it, as shown in this code snippet:

consumer/QueueSizeReporter.cs

```
TagList tags = new () {
    { "net.peer.name", queue.AccountName},
    { "messaging.source.name", queue.Name}};
```

```
_queueSize = _meter.CreateObservableGauge(
    "messaging.azqueues.queue.size",
    () => new Measurement<long>(_currentQueueSize, tags), ...);
```

https://github.com/PacktPublishing/Modern-Distributed-Tracing-in-.NET/blob/main/chapter11/consumer/QueueSizeReporter.cs

We passed a callback that returns a `_currentQueueSize` instance variable. We're going to update it every several seconds as we retrieve the size from the queue:

consumer/QueueSizeReporter.cs

```
var res = await _queue.GetPropertiesAsync(token);
_currentQueueSize = res.Value.ApproximateMessagesCount;
```

https://github.com/PacktPublishing/Modern-Distributed-Tracing-in-.NET/blob/main/chapter11/consumer/QueueSizeReporter.cs

That's it – now we measure the queue size. This number alone does not tell the entire story, but if it's significantly different from the baseline or grows fast, this is a great indication of a problem.

Once the load grows, the queue size will also go up and we may try to add more consumers or optimize them. One of the typical optimizations is batching – it helps reduce the number of network calls and would utilize consumer instances better. Let's see how we can instrument it.

Instrumenting batching scenarios

Instrumentation for batching scenarios can be different depending on the use case – transport-level batching needs a slightly different approach compared to batch processing.

Batching on a transport level

Messages can be batched together to minimize the number of network calls. It can be used by producers or consumers, and systems such as Kafka, Amazon SQS, or Azure Service Bus support batching on both sides.

On the consumer, when multiple messages are received together but processed independently, everything we had for single message processing still applies.

From a tracing perspective, the only thing we'd want to change is to add attributes that record all received message identifiers and batch size on the outer iteration activity.

From the metrics side, we'd also want to measure individual message processing duration, error rate, and throughput. We can track them all by adding a message processing duration histogram.

When we send multiple messages in a batch, we still need to trace these messages independently . To do so, we'd have to create an activity per message and inject unique trace context into each message. The publish activity then should be linked to all the messages being published.

The main question here is when to create a per-message activity and inject context into the message. Essentially, the message trace context should continue the operation that created the message. So, if we buffer messages from different unrelated operations and then send them in a background thread, we should create message activities when the message is created. Then, a batch publish operation will link to independent, unrelated trace contexts.

The duration metric for the publish operation remains the same as for the single-message case we implemented before, but we should consider adding another metric to describe the batch size and the exact number of sent messages – we won't be able to figure it out from the publish duration.

Processing batches

In some cases, we process messages in batches, for example, when aggregating data for analytic purposes, replicating or archiving received data. In such cases, it's just not possible to separate individual messages. Things get even more complicated in scenarios such as routing or sharding, when a received batch is split into several new batches and sent to the next destination.

We can record relationships using links – this will allow us to tell whether (when and how many times) a message was received, and which processing operation it contributed to.

Essentially, we create a batch-processing activity with links to all messages being processed. Links have attributes and there we can put important message metadata, such as the delivery count, message ID, or insertion time.

From a metrics perspective, consumer lag (measured per message), queue size, and processing duration (throughput and failure rate) still apply. We might also want to report the batch size as a histogram.

> **Note**
> Message and batch processing are frequently done outside of messaging client library control, by application code or integration frameworks. It's rarely possible for auto-instrumentation to trace or measure processing calls. These scenarios vary a lot from application to application, requiring custom instrumentations tuned to specific use cases and messaging systems.

Now that we have an idea of how to instrument messaging scenarios, let's see how we can use it in practice.

Performance analysis in messaging scenarios

We're going to use our demo application to simulate a few common problems and use signals we have to detect and debug issues. Let's start the application with the following command:

```
$ docker-compose up --build --scale consumer=3
```

It will run one producer and three consumers along with the observability stack.

You can now send a request to the producer at http://localhost:5051/send, which sends one message to the queue and returns receipt information as a response.

Now you need to add some load with the tool of your choice. If you use bombardier, you can do it with the following command:

```
$ bombardier -c 1 -d 30m http://localhost:5051/send
```

It sends requests to the producer in one connection. You can play with a different number of connections and consumers in the docker-compose command to see how the metrics change.

You might also want to install Grafana and import the dashboard from the book's repository (https://github.com/PacktPublishing/Modern-Distributed-Tracing-in-.NET/blob/main/chapter11/grafana-dashboard.json) to look at all metrics at once.

How do we check whether the consumers are working properly? We can start with consumer lag and queue size metrics. *Figure 11.6* shows the 95th percentile for consumer lag obtained with the following query:

```
histogram_quantile(0.95,
    sum(rate(messaging_azqueues_consumer_lag_seconds_bucket[1m]))
    by (le, messaging_source_name, net_peer_name)
)
```

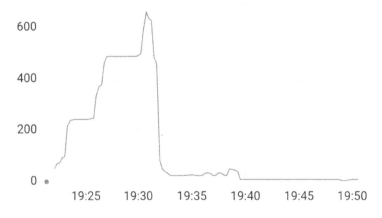

Figure 11.6 – Consumer lag grows over time

Consumer lag grows almost to 600 seconds, and if we look at the queue size, as shown in *Figure 11.7*, we'll see there were up to about 11,000 messages in the queue:

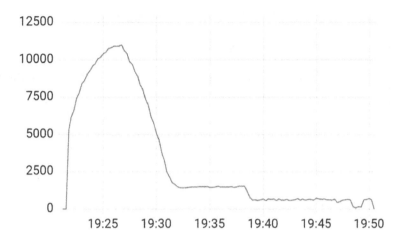

Figure 11.7 – Queue size grows and then slowly goes down

Here's the query for the queue size:

```
max by (net_peer_name, messaging_source_name)
(messaging_azqueues_queue_size_messages)
```

Consumer lag stays high for a long time until all messages are processed at around 19:32, but we can judge by the queue size that things started to improve at 19:27.

The trend changed and the queue quickly shrunk because I stopped the application and restarted it with 15 consumers.

But now we have too many consumers and are wasting resources. We can check the average batch size we retrieve – if it's consistently and noticeably lower than the configured batch size, we may slowly start decreasing the number of consumers, leaving some buffer for bursts of load.

Now, let's stop the load and add some errors. Send a malformed message with `http://localhost:5051/send?malformed=true`. We should see that queue size remains small, but consumer lag grows over time.

We can also see that despite no messages being sent, we're receiving messages, processing them, and failing repeatedly.

For example, we can visualize it with the following query:

```
sum(
  rate(messaging_azqueues_process_loop_duration_milliseconds_
     count[1m]))
by (messaging_source_name, messaging_azqueue_status)
```

It shows the rate of process-and-receive iterations grouped by queue name and status. This is shown in *Figure 11.8*:

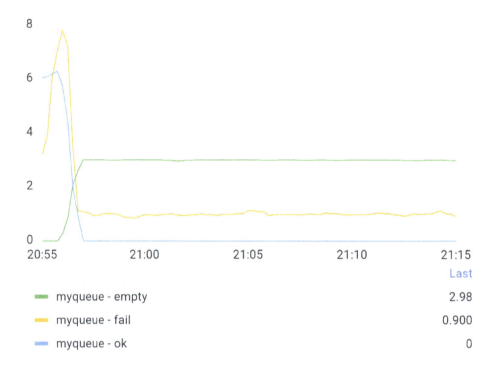

Figure 11.8 – Processing rate grouped by status

We can see here that from around 20:57, we attempt to receive messages about four times per second. Three of these calls don't return any messages, and in the other case, processing fails. There are no successful iterations.

We sent a few malformed messages, and it seems they are being processed forever – this is a bug. If there were more than a few such messages, they would keep consumers busy and not let them process any valid messages.

To confirm this suggestion, let's look at the traces. Let's open Jaeger at `http://localhost:16686` and filter traces with errors that come from consumers. One such trace is shown in *Figure 11.9*:

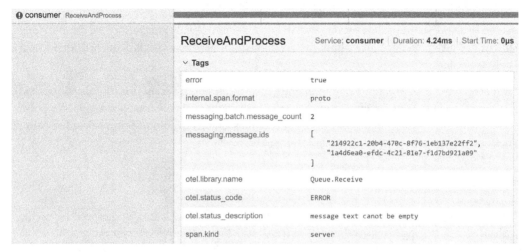

Figure 11.9 – Failed receive-and-process iteration

Here, we see that four messages were received, and the iteration failed with an error. If we could add links to this operation, we would be able to navigate to traces for each individual message. Instead, we have message ID stamped. Let's find the trace for one of these messages using the corresponding attribute. The result is shown in *Figure 11.10*:

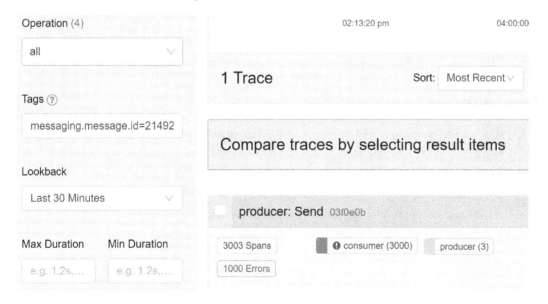

Figure 11.10 – Trace for one of the failed messages

This does not look great – we have 3,000 spans for just one message. If we open the trace and check out the `messaging.azqueues.message.dequeue_count` attribute for the latest processing spans, we'll see the message was received more than 1,000 times.

To fix the issue, we should delete messages that fail validation. We also make sure we do so for any other terminal error and introduce a limit to the number of times a message is dequeued, after which the message is deleted.

We just saw a couple of problems that frequently arise in messaging scenarios (but usually in less obvious ways) and used instrumentation to detect and debug them. As observability vendors improve user experience for links, it will become even easier to do such investigations. But we already have all the means to record telemetry and correlate it in messaging flows.

Summary

In this chapter, we explored messaging instrumentation. We started with messaging specifics and the new challenges they bring to observability. We briefly looked into OpenTelemetry messaging semantic conventions and then dived into producer instrumentation. The producer is responsible for injecting trace context into messages and instrumenting publish operations so that it's possible to trace each independent flow on consumers.

Then, we instrumented the consumer with metrics and traces. We learned how to measure consumer health using queue size and lag and explored the instrumentation options for batching scenarios. Finally, we saw how we can use instrumentation to detect and investigate common messaging issues.

With this, you're prepared to instrument common messaging patterns and can start designing and tuning instrumentation for advanced streaming scenarios.

In the next chapter, we're going to design a comprehensive observability store for databases and caching.

Questions

1. How would you measure end-to-end latency for an asynchronous operation? For example, in a scenario when a user uploads a meme and it takes some time to process and index it before it appears in search results.

2. How would you report batch size as a metric? How it can be used?

3. How would you approach baggage propagation in messaging scenarios

12
Instrumenting Database Calls

In this chapter, we're going to continue exploring instrumentation approaches for popular distributed patterns and will look into database instrumentation. We'll use MongoDB as an example and combine it with Redis cache. We'll add tracing and metrics instrumentation for database and cache calls and discuss how to add application context and provide observability in these composite scenarios. In addition to client-side instrumentation, we'll see how to also scrape Redis server metrics with the OpenTelemetry Collector Finally, we'll explore the generated telemetry and see how it helps with analyzing application performance.

Here's what you'll learn about:

- Tracing MongoDB operations
- Tracing Redis cache and logical calls
- Adding client- and server-side metrics
- Using telemetry to analyze failures and performance

By the end of this chapter, you'll be familiar with generic database instrumentations and will be able to instrument your own applications using databases or caches and analyze their performance.

Technical requirements

The code for this chapter is available in the book's repository on GitHub at `https://github.com/PacktPublishing/Modern-Distributed-Tracing-in-.NET/tree/main/chapter12`.

To run the samples and perform analysis, we'll need the following tools:

- .NET SDK 7.0 or later
- Docker and `docker-compose`

Instrumenting database calls

Databases are used in almost every distributed application. Many databases provide advanced monitoring capabilities on the server side, which include database-specific metrics, logs, or expensive query detection and analysis tools. Client instrumentation complements it by providing observability on the client side of this communication, correlating database operations, and adding application-specific context.

Client instrumentation describes an application's communication with a database ORM system, driver, or client library, which can be quite complicated performing load balancing or batching operations in the background.

In some cases, it could be possible to trace network-level communication between the client library and the database cluster. For example, if a database uses gRPC or HTTP protocols, the corresponding auto-instrumentation would capture transport-level spans. In this case, we would see transport-level spans as children of a logical database operation initiated by the application.

Here, we're going to instrument the logical level of the MongoDB C# driver to demonstrate the principles that apply to other database instrumentations.

> **Note**
> Generic instrumentation for `MongoDB.Driver` is available in the `MongoDB.Driver.Core.Extensions.OpenTelemetry` NuGet package.

Before we start the instrumentation, let's check out OpenTelemetry semantic conventions for databases.

OpenTelemetry semantic conventions for databases

The conventions are available at `https://github.com/open-telemetry/opentelemetry-specification/blob/main/specification/trace/semantic_conventions/database.md`. They have an experimental status and may have changed by the time you access the link.

Conventions define attributes for both logical and physical calls. In our case, we are not instrumenting transport-level communication, so we will only use the ones applicable to logical operations:

- `db.system`: This is a required attribute that tracing backends use to distinguish database spans from all others. It should match the `mongodb` string, which observability backends may use to provide database or even MongoDB-specific analysis and visualizations.

- `db.connection_string`: This is a recommended attribute. It's also recommended to strip credentials before providing it. We're not going to add it to our custom instrumentation. There could be cases where it's useful to capture the connection string (without credentials) as it can help detect configuration issues or we can also log it once at start time.

- db.user: This is yet another recommended attribute that captures user information and is useful to detect configuration and access issues. We're not going to capture it since we have just one user.

- db.name: This is a required attribute defining the database name.

- db.operation: This is a required attribute that captures the name of the operation being executed, which should match the MongoDB command name.

- db.mongodb.collection: This is a required attribute that represents the MongoDB collection name.

In addition to database-specific attributes, we're going to populate MongoDB host information with net.peer.name and net.peer.port – generic network attributes.

Populating network-level attributes on logical calls is not always possible or useful. For example, when a MongoDB driver is configured with multiple hosts, we don't necessarily know which one is used for a particular command. In practice, we should use auto-instrumentation that operates on the command level, subscribing to command events with IEventSubscriber (as described in the MongoDB documentation at http://mongodb.github.io/mongo-csharp-driver/2.11/reference/driver_core/events).

In addition to attributes, semantic conventions require the use of the client kind on spans and providing a low-cardinality span name that includes the operation and database name. We're going to use the {db.operation} {db.name}.{db.mongodb.collection} pattern.

Now that we know what information to include in spans, let's go ahead and instrument a MongoDB operation.

Tracing implementation

In our application, we store records in a MongoDB collection and handle all communication with the collection in a custom DatabaseService class.

Let's start by instrumenting an operation that reads a single record from a collection:

DatabaseService.cs

```
using var act = StartMongoActivity(GetOperation);
try {
  var rec = await _records.Find(r => r.Id == id)
    .SingleOrDefaultAsync();
  ...
  return rec;
} catch (Exception ex) {
  act?.SetStatus(ActivityStatusCode.Error,
```

```
    ex.GetType().Name);
    ...
}
```

Here, we trace the Find method call. We use the GetOperation constant as the operation name, which is set to FindSingleOrDefault – a synthetic name describing what we do here. If the MongoDB command throws an exception, we set the activity status to error.

Let's look in the StartMongoActivity method implementation:

DatabaseService.cs

```
var act = MongoSource.StartActivity(
    $"{operation} {_dbName}.{_collectionName}",
    ActivityKind.Client);
if (act?.IsAllDataRequested != true) return act;

return act.SetTag("db.system", "mongodb")
    .SetTag("db.name", _dbName)
    .SetTag("db.mongodb.collection", _collectionName)
    .SetTag("db.operation", operation)
    .SetTag("net.peer.name", _host)
    .SetTag("net.peer.port", _port);
```

Here, we populate the activity name, kind, and attributes from the semantic conventions mentioned previously. The host, port, database name, and collection name are populated from the MongoDB settings provided via configuration and captured at construction time.

A similar approach could be used for any other operation. For bulk operations, we may consider adding more context to describe individual requests in the array attribute, as shown in this code snippet:

DatabaseService.cs

```
private static void AddBulkAttributes<T>(
    IEnumerable<WriteModel<T>> requests, Activity? act)
{
```

```
  if (act?.IsAllDataRequested == true)
  {
    act.SetTag("db.mongodb.bulk_operations",
      requests.Select(r => r.ModelType).ToArray());
  }
}
```

https://github.com/PacktPublishing/Modern-Distributed-Trac-
ing-in-.NET/blob/main/chapter12/database/DatabaseService.cs

This instrumentation is very generic – it does not record anything application-specific even though it knows the type of the record. For example, we could add a record identifier as an attribute or set the status to error if no records were found. These are all valid things to do if you're going to stick with specialized manual instrumentation, but it's more common to use a shared one when possible.

So, how do we record application-specific context along with generic database instrumentation? One solution would be to enrich auto-collected activities as we did in *Chapter 5, Configuration and Control Plane*.

Another solution is to add another layer of logical activities around database and cache calls. Before we do this, let's learn how to trace cache calls.

Tracing cache calls

Caches such as Redis and Memcached are a special class of databases and are covered by database semantic conventions too. Instrumenting cache calls according to conventions is beneficial as it helps you to stay consistent across all services and to get the most out of your tracing backends in terms of visualization and analysis.

So, let's instrument Redis according to database conventions and add cache-specific context. There is nothing specifically defined in OpenTelemetry for caches, so let's design something of our own.

> **Note**
> Auto-instrumentation for the StackExchange.Redis client is available in the OpenTelemetry.Instrumentation.StackExchangeRedis NuGet package.

When it comes to tracing, we want to know typical things: how long a call took, whether there was an error, and what operation was attempted. Cache-specific things include an indication whether an item was retrieved from the cache or the expiration strategy (if it's conditional) for set operations.

Let's go ahead and instrument a Get call – it looks pretty similar to the database instrumentation we saw in the previous section:

CacheService.cs

```
using var act = StartCacheActivity(GetOperationName);
try
{
  var record = await _cache.GetStringAsync(id);
  act?.SetTag("cache.hit", record != null);
  ...
}
catch (Exception ex)
{
  act?.SetStatus(ActivityStatusCode.Error,
    ex.GetType().Name);
  ...
}
```

https://github.com/PacktPublishing/Modern-Distributed-Tracing-in-.NET/blob/main/chapter12/database/CacheService.cs

Here, we created an activity to trace a GetString call to Redis. If a record is found, we set the cache.hit attribute to true, and if an exception happens, we set the activity status to error and include an exception message.

Let's take a look at the attributes that are set in the StartCacheActivity method:

CacheService.cs

```
var act = RedisSource.StartActivity(operation,
  ActivityKind.Client);
if (act?.IsAllDataRequested != true) return act;
return act.SetTag("db.operation", operation)
    .SetTag("db.system", "redis")
    .SetTagIfNotNull("db.redis.database_index", _dbIndex)
    .SetTagIfNotNull("net.peer.name", _host)
    .SetTagIfNotNull("net.peer.port", _port)
    .SetTagIfNotNull("net.sock.peer.addr", _address)
    .SetTagIfNotNull("net.sock.family", _networkFamily);
```

https://github.com/PacktPublishing/Modern-Distributed-Tracing-in-.NET/blob/main/chapter12/database/CacheService.cs

In this snippet, we start a client activity with the name matching the operation name. We also set all the applicable database and network attributes and add a Redis-specific attribute defined by OpenTelemetry – `db.redis.database_index`. Network attributes, which describe the host, port, IP address, and network family, are populated from Redis configuration options. The `SetTagIfNotNull` method is an extension method defined in our project.

Here, we have the same problem as with MongoDB – Redis configuration options may include multiple servers and we don't know which one is going to be used for a specific call. The instrumentation in the `OpenTelemetry.Instrumentation.StackExchangeRedis` package (we took a quick look at it in *Chapter 3, The .NET Observability Ecosystem*) provides more precise information.

This instrumentation is very generic for the same reasons as for MongoDB – in most cases, we'd rather enrich auto-instrumentation or add another layer of application-specific spans than write a custom instrumentation. So, let's see how we can add the context by adding another layer of instrumentation.

Instrumenting composite calls

With MongoDB and Redis calls instrumented independently and in a generic way, it could be hard to answer questions such as "How long did it take to retrieve a record with a specific ID?" or "How long did retrieval take?" given it involved a call to the cache, a call to the database, and then another call to the cache.

We did not add a record identifier attribute to query on and we only know the duration of individual calls that don't really describe the overall operation.

In the following example, we're adding an extra layer of instrumentation that traces logical operations with a record identifier:

RecordsController.cs

```
using var act = Source.StartActivity("GetRecord");
act?.SetTag("app.record.id", id);
try
{
  var recordStr = await _cache.GetRecord(id);
  if (recordStr != null) return recordStr;

  act?.SetTag("cache.hit", false);
  var record = await _database.Get(id);
  if (record != null) return await Cache(record);
}
catch (Exception ex)
{
  act?.SetStatus(ActivityStatusCode.Error,
```

```
    ex.GetType().Name);
  throw;
}
act?.SetStatus(ActivityStatusCode.Error, "not found");
```

https://github.com/PacktPublishing/Modern-Distributed-Trac-
ing-in-.NET/blob/main/chapter12/database/Controllers/RecordsCon-
troller.cs

Here, we wrap the sequence of calls in the GetRecord activity – it has an internal kind and just two attributes: app.record.id (which captures the record identifier) and cache.hit (describing whether the record was retrieved from the database).

We also provide a not found status description when nothing is found and can report other known issues in the same way.

In the case of our demo application, the encompassing database and cache spans almost match the ASP.NET Core ones in terms of status and duration, but in practice, controller methods do many other things. The encompassing operation helps us separate all spans and logs related to record retrieval.

Now that we have an idea of how to approach tracing, let's explore metrics.

Adding metrics

With databases, it's common to monitor connections and query execution count and duration, contention, and resource utilization in addition to technology-specific things. The MongoDB cluster reports a set of such metrics that you can receive with OpenTelemetry Collector (check it out at https://github.com/open-telemetry/opentelemetry-collector-contrib/tree/main/receiver/mongodbreceiver). These metrics provide the server side of the story. We should also add client-side duration metrics. It'd help us account for connectivity issues and network latency.

OpenTelemetry semantic conventions only document connection metrics for now. We could record them by implementing an IEventSubscriber interface and listening to connection events.

Instead, we're going to record the basic operation duration, which also allows us to derive the throughput and failure rate and slice and dice by operation, database, or collection name.

Let's get back to the Get operation code and see how the metric can be added. First, we'll create a duration histogram:

DatabaseService.cs

```
private static readonly Meter MongoMeter = new("MongoDb");
private readonly Histogram<double> _operationDuration;
...
```

```
public DatabaseService(IOptions<MongoDbSettings> settings) {
  ...
  _operationDuration = MongoMeter.CreateHistogram<double>(
    "db.operation.duration", "ms",
    "Database call duration");
}
```

https://github.com/PacktPublishing/Modern-Distributed-Tracing-in-.NET/blob/main/chapter12/database/DatabaseService.cs

Now that we have a histogram, we can record the duration for each operation:

DatabaseService.cs

```
var start = _operationDuration.Enabled ?
    Stopwatch.StartNew() : null;
using var act = StartMongoActivity(GetOperation);

try
{
  var rec = await _records.Find(r => r.Id == id)
    .SingleOrDefaultAsync();
  TrackDuration(start, GetOperation);
  return rec;
}
catch (Exception ex)
{
  ...
  TrackDuration(start, GetOperation, ex);
  throw;
}
```

https://github.com/PacktPublishing/Modern-Distributed-Tracing-in-.NET/blob/main/chapter12/database/DatabaseService.cs

Here, we call into the TrackDuration method and pass a stopwatch that tracks the duration, the low-cardinality operation name, and an exception (if any). Here's the TrackDuration method:

DatabaseStatus.cs

```
private void TrackDuration(Stopwatch? start,
  string operation, Exception? ex = null)
{
```

```
  if (start == null) return;
  string status = ex?.GetType()?.Name ?? "ok";
  _operationDuration.Record(start.ElapsedMilliseconds,
    new TagList() {
      { "db.name", _dbName },
      { "db.mongodb.collection", _collectionName },
      { "db.system", "mongodb"},
      { "db.operation", operation },
      { "db.mongodb.status", status },
      { "net.peer.name", _host },
      { "net.peer.port", _port }});
}
```

https://github.com/PacktPublishing/Modern-Distributed-Tracing-in-.NET/blob/main/chapter12/database/DatabaseService.cs

Here, we add all the attributes we used for tracing and a new one – db.mongodb.status. We use the exception type as a status to make sure that metric cardinality stays low.

While the idea of using the exception type looks compelling and easy, it only works when we use the same MongoDB driver in the same language across the system. Even then, statuses might change over time with driver updates. In a real production scenario, I would recommend mapping known exceptions to language-agnostic status codes. It also makes sense to test corresponding cases and check that proper error codes are captured. It's important if your alerts are based on specific codes.

The duration histogram and the metrics we can derive from it at query time cover common monitoring needs (throughput, latency, and error rate). We could also use it to do capacity analysis and make better design decisions. For example, before adding a cache in front of the database, we could check the read-to-write ratio to see whether caching would be helpful.

With custom queries over traces, we could also estimate how frequently the same records are accessed. This would help us pick a suitable expiration strategy.

Recording Redis metrics

In addition to common database concerns, we want to measure cache-specific things: the hit-to-miss ratio, key expiration, and the eviction rate. This helps optimize and scale the cache.

These metrics are reported by Redis and can be captured with the Redis receiver for OpenTelemetry Collector, available at https://github.com/open-telemetry/opentelemetry-collector-contrib/tree/main/receiver/redisreceiver.

We can enable them with the following configuration:

configs/otel-collector-config.yml

```yaml
receivers:
  ...
  redis:
    endpoint: "redis:6379"
    collection_interval: 5s


  ...
service:
  pipelines:
    ...
    metrics:
      receivers: [otlp, redis]
  ...
```

https://github.com/PacktPublishing/Modern-Distributed-Trac-ing-in-.NET/blob/main/chapter12/configs/otel-collector-config.yml

OpenTelemetry Collector connects to a Redis instance and scrapes available metrics from it. Redis exposes multiple metrics, including uptime and resource utilization metrics and, most importantly, counters measuring command rate, hits, misses, expirations, evictions, and average time to live. With these, we can monitor Redis' health and see whether it's used efficiently and where the bottlenecks are.

For example, a low hit-to-miss ratio could indicate that we're not utilizing the cache well and potentially could tune caching parameters to make it more efficient. First, we should make sure caching makes sense – usually, it does when at least some items are read more frequently than they are modified. We also need the interval between reads to be relatively low.

If, based on the collected data, we decided to add a cache, we can optimize its configuration further by looking into other cache metrics:

- A high key eviction rate can tell us if we don't have enough memory and keys are evicted before items are read. We might want to scale Redis vertically or horizontally or change the eviction policy to better match the usage pattern. For example, if we have a relatively low number of periodically accessed items, a **least frequently used (LFU)** policy could be more efficient than the **least recently used (LRU)** one.

- If we see a low eviction but high expiration rate, it could mean that the expiration time is too low – items are read less frequently than we expected. We can try to gradually increase the expiration time or disable it and rely on eviction policy instead.

In addition to server-side metrics, we'll also add a client-side duration histogram. It allows us to record call duration distribution with command and other database-specific dimensions. The implementation is almost identical to the MongoDB duration metric. The only difference is that we're going to add the `cache.hit` attribute to the metrics for the `GetString` operation. This could be helpful when server-side metrics are not available or there are multiple different operations we want to measure a hit ratio for independently of each other.

Now that we have all the database traces and metrics in place, let's bring all the pieces together and see how we use this telemetry in practice.

Analyzing performance

Let's first run the demo application using the `$ docker-compose up --build` command. It will start local MongoDB and Redis instances along the application and observability stack.

You can create some records with a tool such as `curl`:

```
$ curl -X POST http://localhost:5051/records \
  -H "Content-Type: application/json" \
  -d '[{"name":"foo"},{"name":"bar"},{"name":"baz"}]'
```

It should return a list of record identifiers the service created.

Now, let's look at the Jaeger trace at `http://localhost:16686`, like the one shown in *Figure 12.1*:

Figure 12.1 – Trace showing bulk record creation

We see a controller span (`Records`) and then `CreateRecords`, which describes a database-and-cache-encompassing operation. It's a parent of the `BulkWrite` span, which describes a MongoDB call and three individual Redis spans – one for each record.

Note that the controller and the `CreateRecords` spans end before caching is complete, because we don't wait for it. Anything that happens within the `SetString` operation would still be properly correlated despite the parent request being complete.

If we were to wait about 10 seconds and try to get one of the records (by calling `http://localhost:5051/records/{id}`), we'd see a trace like the one shown in *Figure 12.2*:

Figure 12.2 – Trace showing record retrieval from the database

If we get the same record within 10 seconds, we'll see it's returned from the cache, as shown in *Figure 12.3*:

Figure 12.3 – Trace showing record retrieval from the cache

By looking at individual traces, we can now quickly see whether records were retrieved from the cache or the database. We can also find all operations that happened across all traces for a specific record using the `app.record.id` attribute or write ad hoc queries using the `cache.hit` flag.

Let's now simulate a failure by stopping the Redis container with `$ docker stop chapter12-redis-1`.

If we try to get one of the records again, the application will return the `500 - Internal Server Error` response. The trace predictably shows that the call to Redis failed with `RedisConnectionException`. We might want to change this behavior, and if the Redis call fails, retrieve the record from the database.

If we did this, we'd see a trace similar to the one shown in *Figure 12.4*:

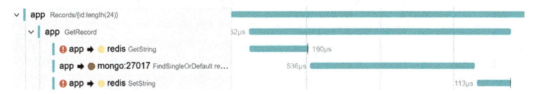

Figure 12.4 – Trace showing Redis call failures with fallback to database

Here, calls to Redis failed, but the overall operation succeeded. You can reproduce it if you comment out the `throw` statement on line 63 in `CacheService.cs` and then rerun the application with `$ docker-compose up --build`.

Let's check what happens with metrics in this case. We can start by applying some load with `loadgenerator$ dotnet run -c Release --rate 50`. Give it a few minutes to stabilize and let's check our application's performance.

Let's first check out the service throughput with the following query in Prometheus (at `http://localhost:9090`):

```
sum by (http_route, http_status_code)
  (rate(http_server_duration_milliseconds_count[1m])
)
```

As we'll see in *Figure 12.6*, throughput stabilizes at around 40-50 requests per second – that's what we configured in the `rate` parameter.

Then, we can check the 50th percentile for latency with the following query:

```
histogram_quantile(0.50,
  sum (rate(http_server_duration_milliseconds_bucket[1m]))
  by (le, http_route, http_method))
```

Later, in *Figure 12.7*, we'll see that responses are blazing fast – the 50th percentile for latency is just a few milliseconds.

> **Spoiler**
>
> If we checked the 95th percentile for latency, we'd notice it is much bigger, reaching 200-300 milliseconds. MongoDB shows these spikes in latency because container resources are constrained for demo purposes.

Let's now check the cache hit rate. We can derive it from Redis server metrics or a client operation duration histogram. The following query uses the latter approach:

```
100 *
sum by (net_peer_name) (
  rate(db_operation_duration_milliseconds_count{cache_hit="true",
        db_operation="GetString",
        db_system="redis"}[1m]))
/
sum by (net_peer_name) (
  rate(
      db_operation_duration_milliseconds_count{db_
        operation="GetString",
      db_redis_status="ok",
      db_system="redis"}[1m]))
```

The query gets the rate of the GetString operation on Redis with the cache.hit attribute set to true and divides it by the overall GetString operation success rate. It also multiplies the ratio by 100 to calculate the hit percentage, which is around 80%, as we can see in *Figure 12.5*:

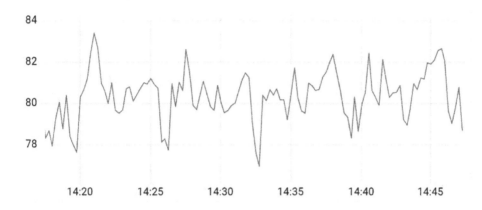

Figure 12.5 – Redis hit rate for the GetString method

So, the cache is used and it handles 80% of read requests. Let's see what happens if we stop it with the $ docker stop chapter12-redis-1 command.

> **Tip**
> With this exercise, you may find it interesting to explore the effect of recording exceptions from Redis. Once the Redis container is stopped, every call to Redis will result in an exception being recorded. In the case of our tiny application, it alone increases the telemetry volume tenfold. Check it out yourself with the following Prometheus query:

```
sum by (container_image_name)
  (rate(container_network_io_usage_rx_bytes_total[1m]))
```

Immediately after the Redis container is stopped (at around 14:48), the application throughput starts to decrease to less than one record per second, as shown in *Figure 12.6*:

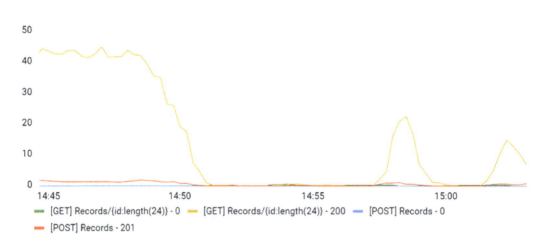

Figure 12.6 – Application throughput before and after the Redis container is stopped

HTTP latency (the 50th percentile) increases from a few milliseconds to several seconds, as you can see in *Figure 12.7*:

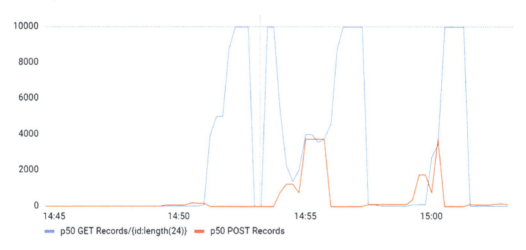

Figure 12.7 – Application latency 50th percentile before and after Redis container is stopped

The spikes in HTTP latency are consistent with the MongoDB latency increase shown in *Figure 12.8*:

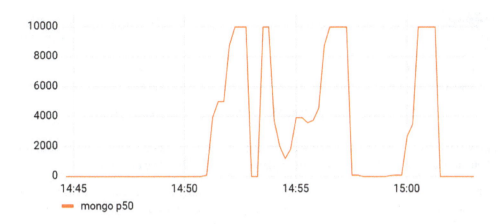

Figure 12.8 – MongoDB latency (p50) in milliseconds

Finally, we should check what happened with MongoDB throughput: since Redis no longer handles 80% of read requests, the load on the database increases and, initially, it tries to catch up, as you can see in *Figure 12.9*:

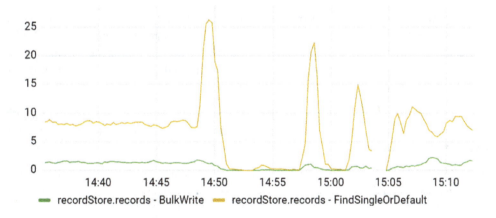

Figure 12.9 – MongoDB throughput before and after the container is stopped

The resources on a MongoDB container are significantly constrained and it can't handle such a load.

If we check the traces, we'll see the MongoDB call takes significantly longer and is the root cause of slow application responses and low throughput. An example of such a trace is shown in *Figure 12.10*:

Figure 12.10 – Trace showing a long MongoDB request when Redis is stopped

If you now start Redis with the $ docker start chapter12-redis-1 command, the throughput and latency will be restored to the original values within a few minutes.

We did this analysis knowing the root cause, but it also works as a general approach – when service-level indicators such as latency and throughput change drastically, we should check the state and health of service dependencies. The findings here are that we need to protect the database better, for example, by adding a few more (potentially smaller) Redis instances that would handle the load if one of them goes down. We may also consider rate-limiting calls to the database on the service side, so it stays responsive, even with lower throughput.

Summary

In this chapter, we explored database instrumentation. We started by looking into OpenTelemetry semantic conventions for databases and implemented tracing for MongoDB. Then, we added similar instrumentation for Redis and encompassing calls. We saw how to provide application-specific context on encompassing spans and record whether data was retrieved from the cache or database to improve performance analysis across traces.

Then, we added metrics, including client duration histograms for MongoDB and Redis along with server-side metrics for Redis that help analyze and optimize cache usage, starting with the hit ratio, which we were able to measure.

Finally, we simulated a Redis outage and saw how collecting telemetry makes it easy to detect and analyze what went wrong and how the outage progressed. We also found several issues in our application that make it unreliable.

Now you're ready to start instrumenting database calls in your application or enrich auto-collected telemetry with additional traces and metrics.

This concludes our journey through instrumentation recipes. In the next chapter, we'll talk about organizational aspects of adopting and evolving tracing and observability.

Questions

1. How would you approach instrumenting a database change feed (the event stream exposed by the database that notifies about changes to database records)? For example, an application can subscribe to a notification that the cloud provider will send when a blob is created, updated, or removed from cloud storage (which we can consider to be a database).

2. Would it make sense to record calls to Redis as events/logs instead of spans?

3. Try removing resource limitations on the MongoDB container and check what happens if we kill Redis now.

Part 4: Implementing Distributed Tracing in Your Organization

This part walks through the sociotechnical aspects of observability adoption – making an initial push and improving it further, developing telemetry standards within your company, and instrumenting new parts of a system in the presence of legacy services.

This part has the following chapters:

13

Driving Change

Throughout the book, we have talked about the technical side of observability and discussed how to trace calls, record metrics, report events, or use auto-collected telemetry provided by platforms and libraries. Here, we're going to talk about the organizational aspects of implementing observability solutions.

First, we'll look into the benefits of and reasons for changing your existing solution and discuss associated costs. Then, we'll go through the implementation stages and come up with a brief. Finally, we'll see how to leverage observability to drive and improve the development process.

in this chapter, you'll learn how to do the following:

- Decide whether you need a better observability solution and which level is right for you

- Develop an onboarding plan and start implementing it

- Use observability to help with daily development tasks

By the end of this chapter, you should be ready to propose an observability solution and onboarding plan for your organization.

Understanding the importance of observability

If you're reading this book, you're probably at least entertaining the idea of improving the observability story of your application. Maybe it's hard to understand how customers use your system or it takes a lot of time to understand what exactly went wrong when someone reports an issue. In the worst case, it takes a lot of time to just notice that the system is unhealthy and users are affected. Or, maybe you want to minimize such risks in your future projects.

In any case, these pain points brought you here and they should guide you further to find the right observability level and approach for your system.

Even if we clearly see the problem and how it can be solved with better observability, we usually still need to get other people working on the system onboard with this vision. Astoundingly, they might have quite different feelings about the same problems and might not consider them worthy of solving.

Let me share a few common points I have heard arguing that a problem is not important:

- When a customer reports an issue, we can ask for a timestamp and find operations at that time by customer identifier. Then we can find any suspicious logs, get the request ID, and then find correlated logs on other services.

- When we see an issue in production, we open related dashboards and start visually correlating metrics until we can guess what's broken and then mitigate it. We have experts and an excellent set of runbooks for typical issues.

- We can do a user study or customer research to get extensive information on how people use the system.

- We ask customers to enable verbose logs and reproduce the problem, then send us logs that we'll parse based on our expert knowledge of the system.

> **Note**
> Each of these approaches is totally valid. They *already* solve the problem, your team *already* knows how to use them, and some of them are still necessary and quite useful even with perfect observability solutions in place.

So, essentially, when we consider the approach to observability, we need to break the status quo and convince ourselves and our organization that it's worth it. To achieve this, we first need to clearly outline the pain points and understand the cost of keeping things as they are.

The cost of insufficient observability

Your organization might already have some common incident metrics in place that we can rely on, such as **MTTM (mean time to mitigate)**, **MTTR (mean time to recover)**, **MTBF (mean time between failures)**, or others. They are somewhat subjective and depend on what qualifies as an incident, or what recovery means, but roughly show how fast we can investigate incidents and how frequently they happen.

If incidents take a lot of time to resolve and happen frequently, it's likely that our organization cares deeply about them and would be interested in improving the situation.

Ironically, we need at least some level of observability to notice there is an incident and to measure how long it takes to resolve. If we don't have even this in place, we can start to manually track when things get broken and how long it takes us to discover and resolve issues. However subjective it is, it's better than nothing.

Some things rarely appear in such metrics directly: how bad is your on-call experience? How much time does onboarding take before someone can be on-call independently? How many issues end up closed with something such as "cannot reproduce," "not enough information," "probably a network or hardware error"; get lost in ping-pong between teams; or get moved to backlogs and never resolved?

It should be feasible to measure some of such things. For example, we can label issues that can't be investigated further due to a lack of telemetry. If they represent a significant portion of your bugs, it's something worth improving.

As a team, you can also do an experiment for a week or two to roughly measure the time spent investigating issues. How much time does it take to investigate when there is enough data? Or, how much time is wasted investigating issues and meeting a dead end due to a lack of telemetry or finding a trivial transient network issue?

By minimizing the time necessary to find the root cause of an issue, we improve the user experience. We notice incidents earlier and resolve them faster. We also improve our work-life balance and focus on creative work instead of grepping megabytes of logs.

> **Note**
>
> There could be other data, such as business analytics, support stats, public reviews, or anything else, showing that a noticeable number of users are leaving us because of unresolved technical issues. If you need to convince your organization to invest in observability, finding such data and showing how a better observability story can improve things could be a good way to approach it.

So, the first step is to understand whether current tools and processes are effective and have a rough understanding of how better observability could improve things. The next step is to understand the cost of the solution.

The cost of an observability solution

We can roughly break down the costs into two groups: implementation and telemetry backend costs.

We need to add instrumentation, tune and customize telemetry collection, learn how to use new tools, create alerts and dashboards, and build new processes around them. When onboarding a mature and stable system, we should also consider risks – we might break something and temporarily make it less reliable.

As we discussed in *Chapter 9, Best Practices*, we can always choose the level of detail and amount of customization to help us keep the costs within the given budget.

The minimalistic approach would be to start with network-level auto-instrumentation for actively developed services and then add context, customizations, and manual instrumentation as we go.

By using OpenTelemetry and shared instrumentation libraries, we can also rely on vendors to provide common visualizations, alerts, dashboards, queries, and analysis for typical technologies. As a result, it's almost free to get started.

Telemetry backend

We can host the observability stack ourselves or use one of the available platforms. Either way, there will be recurring costs associated with using the solution.

These costs depend on telemetry volume, retention period, the number of services and instances, and many other factors, including the support plan.

Throughout the book, we have discussed how to optimize telemetry collection while keeping the system observable enough for our needs: traces can be sampled, metrics should have low cardinality, and events and logs can be sampled too or kept in cold but indexed storage.

It's a good idea to try a few different backends – luckily, many platforms have a free tier or trial period and, most importantly, you can instrument the system once with OpenTelemetry and pump data into multiple backends to compare the experience and get an idea of what the cost of using them would look like. Once you start relying on a specific backend for alerts or daily tasks, it will be more difficult to switch between vendors.

During this experiment, you will also get a better understanding of the necessary data retention period, sampling rate, and other parameters, and will be able to pick them along with the vendor.

> **Note**
> When running a modern cloud application under scale, it's not possible to operate it without an observability solution, so it's not a question of whether you need one but rather how many details you need to collect and which observability vendor out there works best for your system and budget.

Essentially, we can start small and incrementally tune collection to add or remove details, while keeping it within a reasonable budget.

We should also define what success means – it could be an MTTR improvement, subjective user experience, on-call engineer happiness, anything else that matters for your organization, or any combination of these.

Let's now talk more about the implementation details and try to make this journey less painful.

The onboarding process

The need for distributed tracing and visibility of all parts of the system comes from the complexity of modern applications. For example, we need to know how a serverless environment interacts with cloud storage to debug configuration issues or optimize performance. Or, maybe we want to know why certain requests fail in the downstream service without asking someone to help.

To make the most of distributed tracing, we have to onboard the whole system (or at least a significant part of it), making sure all services create correlated and coherent telemetry, write it to a place where different teams can access it, and reuse the same tooling to do analysis.

So, implementing an observability solution is an organization-wide effort, and it makes sense to start with a pilot project instrumenting a small part of the system. Let's outline its scope and goals.

The pilot phase

The goal of this project is to get hands-on experience with observability, discover any significant technical or process issues early, and better understand the scope and effort needed for the rest of the system.

We'll need a few (at least two) services to start instrumenting:

- That are in active development
- That interact with each other
- That have no (or few) internal dependencies except on each other
- That have teams working on them in close contact

From the technical side, we'll use this phase to make the following decisions:

- **Instrumentation SDKs**: I hope I convinced you to use .NET platform capabilities and OpenTelemetry, but you probably need to decide what to do with existing instrumentation code and tooling.

- **Context propagation standards**: Using W3C Trace Context would be a good start. We may also need to decide whether and how to propagate baggage, or how to pass context over non-HTTP/proprietary protocols.

- **Sampling strategy**: Rate-based, percentage-based, parent-based, tail-based – these are good things to decide early on and identify whether you need an OpenTelemetry collector or can rely on an observability vendor.

- **Which vendor to use and a migration plan from your current one**: We'll discuss technical aspects and trade-offs when instrumenting existing systems in *Chapter 15, Instrumenting Brownfield Applications*.

By the end of the pilot phase, we should have a clear understanding of what onboarding takes, what the challenges are, and how we will solve them.

We will also have a small part of the system instrumented – it's a good time to check whether we see any improvement.

Tracking progress

In the perfect world, after instrumentation is deployed, we'd be able to resolve all incidents in no time and investigate all the tricky bugs we've hunted down for months. I wish that was the case.

There are at least several challenges along the way:

- **Change is hard**: People prefer to use tools they know well, especially when they're dealing with incidents in production. It would be a good exercise to do the same investigation with the new observability solution after the incident is resolved and compare the experiences.

  It's best to start playing with new tools at development time or when investigating low-priority failures. In any case, it takes time and practice to learn about and trust new tools.

- **You'll discover new issues**: Looking at traces or a service map for the first time, I always learn something new about my code. It's common to discover calls to external systems you didn't expect (for example, auth calls made under the hood), unnecessary network calls, wrong retry logic, calls that should run in parallel but run sequentially, and so on.

- **Basic auto-instrumentation is not sufficient**: Without application context or manual instrumentation for certain libraries and scenarios, our ability to find, understand, and aggregate related telemetry is limited.

It will take a few iterations to see an improvement – make sure to collect feedback and understand what's working and what's not.

It also takes time and dedication. Demos, success stories, shared case studies, and documentation on how to get started should create awareness and help people get curious and learn faster.

Iterating

So, after the initial instrumentation, we're not quite ready to roll it out to the rest of the system. Here're a few things to do first:

- **Tune instrumentation libraries**: Remove verbose and noisy signals or enable useful attributes that are off by default. If some parts of your stack don't have auto-instrumentation available, start writing your own.

- **Add essential application context**: Finding common properties and standardizing attribute names or baggage keys across your organization will have a huge impact down the road. We'll talk more about it in *Chapter 14, Creating Your Own Conventions*.

- **Start building backend-specific tooling**: Alerts, dashboards, and workbooks will help us validate whether we have enough context and telemetry to run our system and migrate to the new solution.

> **Note**
>
> By the end of this stage, you should be able to see positive outcomes. It probably will not yet have moved the needle for the whole system, and there might not be enough data for the services involved in the experiment, but you should at least see some success stories and be able to show examples of where the new solution shone.

If you see cases where the new observability story should have helped but has not, it is a good idea to investigate why and tune instrumentation further. While iterating, it's also worth paying attention to backend costs and optimizing telemetry collection if you see the potential for a significant reduction without noticeable impact.

The goal here is to create a good enough instrumentation approach and any necessary tooling around it. We iterate fast and keep the number of participating services small so we can still change direction and make breaking changes.

Once we have finalized all the decisions and implemented and validated them on a small part of the system, we should be able to rely on new observability solutions for most of the monitoring and debugging needs. Before we roll them out, we still need to document them and create reusable artifacts.

Documenting and standardizing

The main outcome of the pilot phase is clarity on how to make the rest of the system observable and the specific benefits it will bring.

To maximize the impact of this phase, we need to make it easier for other services to be onboarded. We can help them by doing the following:

- Documenting new solutions and processes
- Providing demos and starters showing how to use backends and configure them, and add alerts and dashboards
- Producing common artifacts that include any customizations:
  - Context propagators, samplers, or instrumentations
  - Attribute names or helpers that efficiently populate them
  - Starter packs that bring all OpenTelemetry dependencies and enable telemetry collection in a uniform manner
  - Common configuration options

Finally, we're ready to instrument and onboard the rest of the system. It will probably take some time to align attribute names, configuration, or backend plans. We also need to keep tracking progress toward original goals and apply necessary changes when things don't work. Let's talk about a few things that can slow us down.

Avoiding pitfalls

The challenge with distributed tracing and observability is that they're most effective when a distributed application produces coherent signals: trace context is propagated, sampling algorithms are aligned to produce full traces, all signals use the same attribute names, and so on.

While OpenTelemetry solves most of these concerns, it still relies on the application to bring all the pieces together and use coherent signals. It becomes a problem for huge organizations where one service deviating from the standard breaks the correlation for the rest of the system.

Here are a few things to avoid when onboarding your system:

- **Starting too big**: If multiple teams work on instrumentation independently, they will inevitably develop different solutions optimized for their services. Aligning these solutions after onboarding is finalized would be a difficult project on its own. Each team would have an impression that things work for them, but end-to-end customer issues would still take months to resolve.

- **Not sharing telemetry across the system**: When investigating issues or analyzing performance and usage, it's beneficial to be able to see how other services process requests. Without doing so, we will end up in the same place where each cross-service problem involves some amount of ping-pong and problems are not resolved fast enough.

- **Not enforcing standards**: Inconsistent telemetry would make us come back to grepping logs and make customers and ourselves unhappy.

- **Not using new tools and capabilities**: We talked about how migrating from familiar tooling is hard. We need to put enough effort into advocating, promoting, explaining, documenting, and improving things to make sure people use them. Sunsetting old tools, once new ones are more capable and fast, is one (sometimes unpopular) way to make sure everyone switches.

- **Not using the observability stack at development or test time**: Investigating flaky tests is one of the cheapest ways to debug tricky issues. Traces can help a lot, so make sure tests send traces and logs by default and it's super easy to enable tracing on dev machines.

- **Building things that are hard to use or not reliable**: While some friction is expected, we should make sure most of it happens during the pilot phase. If you decide to build your own observability stack based on OSS solutions, you should expect that certain things such as navigating across tools will be difficult and you'll need to put a decent amount of effort into making them usable. Another big investment is building and maintaining reliable telemetry pipelines.

Hopefully, we can avoid most of these issues and onboard a significant part of the system onto our new observability stack.

As we continue onboarding, we should see improvement toward our initial goals and may need to adjust them. During the process, we probably learned a lot about our system and are now dealing with new challenges we could not see before due to a lack of observability.

The journey does not end here. In the same way that we never stop writing tests, we should incorporate and leverage observability in day-to-day tasks.

Continuous observability

Observability should not be added as an afterthought when service or feature development is over. When implementing a complex feature across several services or just adding a new external dependency, we can't rely on users telling us when it's broken. Tests usually don't cover every aspect and don't represent user behavior.

If we don't have a reliable telemetry signal, we can't say whether the feature works or whether customers use it.

Incorporating observability into the design process

Making sure we have telemetry in place is part of feature design work. The main questions the telemetry should answer are the following:

- Who uses this feature and how much?

- Does it work? Does it break something else?

- Does it work as expected? Does it improve things as expected?

If we can rely on the existing telemetry to answer these questions, awesome!

We should design instrumentation in a way that covers multiple things at once. For example, when we switch from one external HTTP dependency to a new one, we can leverage existing auto-collected traces and metrics. A common processor that stamps application context on all spans will take care of traces from the new dependency as well.

If we use feature flags, we should make sure we record them on telemetry for operations that participate in an experiment. We can record them on events or spans, for example, following OpenTelemetry semantic conventions for feature flags available at `https://github.com/open-telemetry/opentelemetry-specification/blob/main/specification/trace/semantic_conventions/feature-flags.md`.

In some cases, default telemetry is not sufficient and we need to add custom events, traces, metrics, or at least extra attributes. It's rarely a good idea to limit instrumentation to a new log record unless we write it in a structured and aggregable way.

Once a feature is proven useful and fully rolled out, we might want to remove this additional telemetry along with the feature flag. It's a great approach if we're sure it's not necessary anymore. Cleaning up and iterating on instrumentation is another important aspect.

Housekeeping

As with any other code, instrumentation code degrades and becomes less useful when neglected.

Similar to the test code, any observability code we write is less reliable than application code – it's also hard to notice when it reports something incorrect as there is no functional issue. So, validating it with testing or manual checks and fixing it in a timely manner is important.

This is one of the reasons to use popular instrumentation libraries – they have been through excessive testing by other people. Keeping your instrumentation libraries up to date and sharing custom ones across the company (or open sourcing them) will result in better instrumentation quality.

Another important part is to make small improvements as you notice issues: add missing events, spans, and attributes (don't forget to check if there is a common one), structure and optimize logs, and adjust their verbosity.

These changes might be risky. We might remove something that people rely on for alerting or analysis. There could be some additional guards in place that prevent it – code reviews, documentation, and tests, but it is rarely possible to account for everything, so be cautious when removing or renaming things.

Another risk is adding something expensive or verbose that would either impact application availability, overwhelm telemetry pipelines, or significantly increase your observability bill. Paying attention to the dev and test telemetry and knowing what's on the hot path should prevent obvious mistakes.

Building reliable telemetry pipelines with rate-limiting should decrease the severity of such incidents when they make it to production.

As you can see, observability code is not very different from any other piece of infrastructure. Implementing it starts with some research and experiments and works best when we tune and improve it along with our application.

Summary

In this chapter, we discussed how to implement and roll out observability solutions in your organization. These efforts can be motivated and justified by the current monitoring infrastructure not being efficient in investigating and resolving customer issues.

We discussed how we can rely on existing metrics or data to understand whether there is room for improvement and estimate the cost of inaction. Then we looked into common costs associated with implementing and running a modern observability solution – the easiest way to find out is to run a small experiment and compare different vendors.

We explored how we can approach onboarding by starting with a pilot project on a small part of the system and iterating and validating results before we roll it out to the rest of the system. Finally, we discussed the importance of incorporating observability into daily tasks and evolving it along with the code.

This chapter should help you justify initial observability investments and gradually implement the solution across the system. In the next chapter, we'll talk more about unifying telemetry collection and introducing your own standards.

Questions

1. Should we look for a single backend for all telemetry signals or a combination of them optimized for individual telemetry signals?

2. How would you approach standardizing baggage propagation and usage in your system?

3. You're adding a cache to the service. When would you add instrumentation? How would you approach it?

Further reading

- *Becoming a Rockstar SRE* by Jeremy Proffitt and Rod Anami

14

Creating Your Own Conventions

Correlation is one of the most important parts of observability. Distributed tracing bring correlation by propagating trace context, allowing us to follow individual operations, and consistent attributes enable correlation across traces and other telemetry signals.

In *Chapter 9, Best Practices*, we talked about the importance of reusing standard attributes and following OpenTelemetry semantic conventions. Sometimes we need to go further and define our own conventions. Here, we're going to explore how to define custom attributes and conventions and use them across the system.

First, we'll list properties that should be standardized across the system, and then we'll explore how to populate them with shared code. Finally, we'll look at OpenTelemetry's semantic convention schema and see how it can simplify documenting and validating custom conventions.

In this chapter, you'll learn how to do the following:

- Identify and document common attributes and conventions
- Share instrumentation and custom conventions across the system
- Use OpenTelemetry tooling to create conventions

With this, you should be able to create easy-to-use processes and tools to keep custom telemetry and attributes consistent and stable.

Technical requirements

The code for this chapter is available in the book's repository on GitHub at `https://github.com/PacktPublishing/Modern-Distributed-Tracing-in-.NET/tree/main/chapter14`.

To run the examples and perform analysis, we'll need the following tools:

- .NET SDK 7.0 or later
- Docker and `docker-compose`

Defining custom conventions

There are multiple ways to express even the most basic things. If we take our meme application example from *Chapter 5*, *Configuration and Control Plane*, we enriched all spans with the meme name attribute so that we can find when the meme was uploaded or how frequently it's accessed.

We chose that approach, but we could instead write a log with the meme name once, and then use slightly more complicated queries to find all traces related to that meme. We could come up with something else, but what's important is to keep the approach consistent across the system.

Even with a custom attribute added to each span, there are still plenty of things to consider when recording such an attribute:

- **Attribute name**: `meme_name`, `meme.name`, and `memeName` are different attributes. Unless we document the exact name and define it as a constant somewhere, someone will eventually use the wrong variation of it.

- **Type**: The meme name is just a string. What if we wanted to capture the size or the image format? We need to document the type and maybe provide helper methods to record attributes so that it's easier to set them correctly.

- **Value**: We need to document what this attribute represents. For example, if we add the image format, should it be represented with a MIME type such as `image/png` or as an enumeration?

  When recording a meme name, we extract it from the uploaded filename, and we need to document what should be recorded on the attribute value (absolute or relative path, filename, with or without extension). If, when writing business logic, we sanitize or escape the meme name, or generate a unique one, we probably want to capture the one used by the business logic. We may also record the original name just once for debugging purposes.

- **When to populate the attribute**: Document on which metrics, spans, and logs the attribute should be recorded. For example, meme names have high cardinality and do not belong on metrics. The meme name attribute can be recorded on spans and logs, but which ones? In our example in *Chapter 5*, *Configuration and Control Plane*, we recorded the meme name on all spans and a couple of specific log records.

There are other aspects you might want to document: relationships between spans, event names, whether to record exceptions on spans, attribute cardinality, stability, and so on.

We'll see how to formally define attributes in the *Using OpenTelemetry schemas and tools* section of this chapter. Now, let's focus on naming.

Naming attributes

Naming is known to be one of the hardest problems in computer science.

If we named the meme name attribute `document.id`, it might precisely match a property in the database schema. However, it would be very generic and have a high chance of collision with another similar concept in the system. People who are not familiar with the internals but analyze business data might have a hard time understanding what `document.id` represents.

`meme_name` seems to be intuitive, short, and descriptive, and the chances of a collision with something else are low. It seems fine, but we'll probably have other attributes and should put them into an application-specific or company-specific namespace.

Namespaces

Namespaces allow us to set unique, specific, descriptive, and consistent names.

Since we have called our system memes, let's use it as the root namespace. This will help us understand that all attributes in this space come from our customizations and are not set by some auto-instrumentation.

This helps us navigate between telemetry and makes it possible to do filtering, redaction, and other post-processing in telemetry pipelines. For example, if we want to remove unknown attributes from logs, we can consider everything in the memes namespace (`http`, `db`, and so on) to be known.

We can have nested namespaces. Since we'd like to record other meme properties, such as size and type, we can end up with the following set of attributes: `memes.meme.name`, `memes.meme.size`, and `memes.meme.type`. We can easily add other properties as we see fit, such as `memes.meme.author` or `memes.meme.description`.

While `memes.meme` might look repetitive, it'll start to make more sense once we add something such as `memes.user.name` or `memes.tag.description`.

The OpenTelemetry naming convention (available at `https://github.com/open-telemetry/opentelemetry-specification/blob/main/specification/common/attribute-naming.md`) uses a dot (`.`) as a separator between namespaces.

For multi-word namespaces or attributes, OpenTelemetry recommends `snake_case`: for example, we could introduce `memes.meme.creation_date`.

Following this convention for our custom attributes allows us to stay consistent across all attributes. It will also reduce the chance of mistakes while writing queries.

Having a schema defined and documented in some form is an essential step. But how can we keep it in sync with the code and ensure all services follow it? One way to do so is by capturing it in code and reusing it across the system.

Sharing common schema and code

Consistent telemetry reporting applies to telemetry collection configuration. First, we need to enable a basic layer of instrumentation on all services, which should include resource utilization metrics, traces, and metrics for HTTP, gRPC, or any other RPC protocol used in your system.

We should also configure sampling and resource attributes, add enrichment processors, and set up context propagators.

Individual services should be able to customize configuration to some extent: add more instrumentations, enable custom activity sources and meters, or control log verbosity.

The easiest way to unify configuration is to ship corresponding code as a common library (or a set of them) shared across all the services in your system. Such libraries would define configuration options, provide helper methods to enable telemetry collection, implement common enrichment processors, declare cross-service events, and so on. Let's go ahead and implement such a configuration helper.

Sharing setup code

In *Chapter 5*, *Configuration and Control Plane*, and other chapters where we used meme application, we applied OpenTelemetry configuration individually in each service.

We'd never do this in production code – it's hard to keep our configurations, instrumentation options, OpenTelemetry package versions, and anything else in sync.

To fix this, we can start extracting common pieces of instrumentation code into a shared library. Since the configuration can slightly vary from service to service, we'll need to define some configuration options.

We need are options that would help us to set the service name, specify the sampling rate or strategy, enable additional instrumentations, and more. You can find an example of such options in the `MemesTelemetryConfiguration` class in the book's repository.

Then we can declare a helper method that takes care of the OpenTelemetry configuration. Here's an example of this method:

OpenTelemetryExtensions.cs

```
public static void ConfigureTelemetry(
  this WebApplicationBuilder builder,
  MemesTelemetryConfiguration config)
{
  var resourceBuilder = GetResourceBuilder(config);
  var sampler = GetSampler(config.SamplingStrategy,
    config.SamplingProbability);

  builder.Services.AddOpenTelemetry()
```

```
    .WithTracing(builder => builder
      .SetSampler(sampler)
      .AddProcessor<MemeNameEnrichingProcessor>()
      .SetResourceBuilder(resourceBuilder)
      .AddHttpClientInstrumentation(o =>
        o.ConfigureHttpClientCollection(
          config.RecordHttpExceptions))
      .AddAspNetCoreInstrumentation(o =>
        o.ConfigureAspNetCoreCollection(
          config.RecordHttpExceptions,
          config.AspNetCoreRequestFilter))
      .AddCustomInstrumentations(config.ConfigureTracing)
      .AddOtlpExporter())
  ...
}
```

https://github.com/PacktPublishing/Modern-Distributed-Tracing-in-.NET/blob/main/chapter14/Memes.OpenTelemetry.Common/OpenTelemetryExtensions.cs

All services that use this method will have the same basic level of instrumentation applied and enriched consistently.

Here's an example of using this method in the **storage** service:

storage/Program.cs

```
var config = new MemesTelemetryConfiguration();

builder.Configuration.GetSection("Telemetry").Bind(config);
config.ConfigureTracing = o => o
  .AddEntityFrameworkCoreInstrumentation();

builder.ConfigureTelemetry(config);
```

https://github.com/PacktPublishing/Modern-Distributed-Tracing-in-.NET/blob/main/chapter14/storage/Program.cs

Here, we read the telemetry options from the `Telemetry` section of the ASP.NET Core configuration, which you can populate in any way that works for you.

Then, we add the Entity Framework instrumentation. Only the **storage** service needs it.

> **Note**
>
> Having a central library that enables collection helps to reduce version hell. By having a dependency on it, individual service packages get a transitive dependency on OpenTelemetry packages and should never add them as direct dependencies. Services that need an uncommon instrumentation library would still need to install the corresponding NuGet package.

Now that we have a common setup, let's see what we can do to help services follow our custom semantic conventions.

Codifying conventions

In the previous example, we started recording the meme name attribute – we enabled `MemeNameEnrichingProcessor`, which sets the `memes.meme.name` attribute on each span. Individual services don't need to do anything to enable it and cannot set the wrong attribute name.

Still, we might need to use the attribute directly in some other parts (for example, on logs), so it's important to declare the attribute name as a constant and never use a string literal in the code. Here's an example demonstrating how attribute names can be declared:

SemanticConventions.cs

```
public class SemanticConventions
{
    public const string MemeNameKey = "memes.meme.name";
    public const string MemeSizeKey = "memes.meme.size";
    public const string MemeTypeKey = "memes.meme.type";
    ...
}
```

https://github.com/PacktPublishing/Modern-Distributed-Tracing-in-.NET/blob/main/chapter14/Memes.OpenTelemetry.Common/SemanticConventions.cs

OpenTelemetry also provides the `OpenTelemetry.SemanticConventions` NuGet package, which declares common attributes defined in the specification. It might make sense to add a dependency to it when reusing common attributes.

So, we have now defined constants for the attribute names, and we have a processor that populates the meme name. Can we do more?

We can provide helpers to report common events in a performant and consistent manner. Let's look at how we can use the high-performance logging we explored in *Chapter 8, Writing Structured and Correlated Logs*, to populate our attributes:

EventService.cs

```
private static readonly Action<ILogger, string, string?,
  long?, string, string, Exception> LogDownload =
    LoggerMessage.Define<string, string?,
      long?, string, string>(
    LogLevel.Information,
    new EventId(1),
    $"download {{{SemanticConventions.MemeNameKey}}}
    {{{SemanticConventions.MemeTypeKey}}}
    {{{SemanticConventions.MemeSizeKey}}}
    {{{SemanticConventions.EventNameKey}}}
    {{{SemanticConventions.EventDomainKey}}}");
  ...
  public void DownloadMemeEvent(string memeName,
    MemeContentType type,
    long? memeSize) =>
  LogDownload(_logger,
    memeName,
    ContentTypeToString(type),
    memeSize,
    SemanticConventions.DownloadMemeEventName,
    SemanticConventions.MemesEventDomain,
    default!);
```

https://github.com/PacktPublishing/Modern-Distributed-Trac-ing-in-.NET/blob/main/chapter14/Memes.OpenTelemetry.Common/Event-Service.cs

Here, we have some difficult-to-read code. It defines a log record representing the meme download event by following the OpenTelemetry conventions and our own conventions. We would not want to write this code every time we needed to log something.

Implementing this event once in the common library and making it easy to reuse is the best way to record the event consistently and with a low performance overhead.

Now, anyone can use the `DownloadMemeEvent` method, as shown in the following example:

StorageService.cs

```
_events.DownloadMemeEvent(name, MemeContentType.PNG,
    response.Content.Headers.ContentLength);
```

https://github.com/PacktPublishing/Modern-Distributed-Tracing-in-.NET/blob/main/chapter14/frontend/StorageService.cs

This is easy to use and performant, and there is no need to worry about attributes, their types, or any conventions at all. If attributes are renamed, there is no need to update the service code – it's all hidden in the shared library.

If we follow this approach, we can define other events and add helper methods to populate attribute groups on metrics and traces.

If we need any custom instrumentations, like we had for gRPC or messaging, we should put them into shared libraries and apply all the attributes there instead of in the service code.

Separating telemetry-related code from business logic makes them both easier to read and maintain. It also makes telemetry-related code testable and helps us to keep it in sync with documentation. It also becomes easy to enforce conventions with tests and notice during code review when shared code changes something controlled by semantic conventions and breaks them.

Defining semantic conventions in code is sufficient for some applications. Others, which may use different languages or have some other constraints, cannot rely on shared code alone. Either way, telemetry can be used by everyone in the company for business reporting and for any non-technical needs. So, it's important to document it separately from code.

Let's see how we can do this using OpenTelemetry tooling.

Using OpenTelemetry schemas and tools

It does not really matter how we document custom semantic conventions. The goal is to have a consistent and specific convention that's easy to read and follow. Let's see how the OpenTelemetry semantic conventions schema may help with this.

Semantic conventions schema

So far, when we have talked about semantic conventions, we have referred to Markdown files such as https://github.com/open-telemetry/opentelemetry-specification/blob/main/specification/trace/semantic_conventions/http.md. These files are the source of truth, but here, we're going to take a look at the implementation details behind them.

The tables describing the attributes in these files are usually auto-generated. Attributes are defined in YAML files that follow OpenTelemetry's semantic convention schema.

YAML files could be shared across different semantic conventions and signals and then consistently written to all Markdown files with a script.

Let's see how our meme attributes can be defined in a YAML file to get an impression of the schema:

memes-common.yaml

```
groups:
  - id: memes.meme
    type: attribute_group
    brief: "Describes memes attributes."
    prefix: memes.meme
    attributes:
      - id: name
        type: string
        requirement_level: required
        brief: 'Unique and sanitized meme name'
        examples: ["this is fine"]
```

https://github.com/PacktPublishing/Modern-Distributed-Trac-ing-in-.NET/blob/main/chapter14/semconv/memes-common.yaml

Here, we defined the name attribute with a string type inside the memes.meme namespace (defined by the prefix property). It's a required attribute since, in *Chapter 5, Configuration and Control Plane*, we decided to record the meme name attribute on all the spans.

OpenTelemetry supports several requirement levels:

- required: Any telemetry item that follows this convention must set the attribute.

- conditionally_required: The attribute must be populated when a condition is met. For example, http.route is only populated when routing is enabled and a route is available.

- recommended: The attribute should be populated, but may be removed or disabled. Observability backends and tools should not rely on it being available. This is the default level.

- opt_in: The attribute is not populated by default, but is known and documented, and it can be added when it's explicitly enabled.

Let's see how we can define the `size` attribute:

---

memes-common.yaml

```
- id: size
  type: int
  requirement_level: opt_in
  brief: 'Meme size in bytes.'
  examples: [49335, 12345]
```

https://github.com/PacktPublishing/Modern-Distributed-Trac-ing-in-.NET/blob/main/chapter14/semconv/memes-common.yaml

---

The `size` attribute has the `int` type (and maps to `int64` or `long`) and an `opt-in` level as we don't want to record it on all telemetry by default.

Finally, we can define the `type` attribute:

---

memes-common.yaml

```
- id: type
  type:
  members:
    - id: png
      value: "png"
      brief: 'PNG image type.'
    - id: jpg
      value: "jpg"
      brief: 'JPG image type.'
    - id: unknown
      value: "unknown"
      brief: 'unknown type.'
  requirement_level: opt_in
  brief: 'Meme image type.'
  examples: ['png', 'jpg']
```

https://github.com/PacktPublishing/Modern-Distributed-Trac-ing-in-.NET/blob/main/chapter14/semconv/memes-common.yaml

---

Here, we define `type` as an enumeration. Instrumentation must use one of the values defined here or set `type` to `unknown`.

I hope your team won't spend too much time deciding between JPEG and JPG – either is fine. What's important is to pick and document one option.

You can find the full schema definition in the OpenTelemetry build tools repository on GitHub (`https://github.com/open-telemetry/build-tools/blob/main/semantic-conventions/syntax.md`). It also contains the schema definition that your IDE may use to autocomplete and validate schema files.

Now that we have defined a few attributes, let's use another OpenTelemetry tool to validate the schema file and generate content in a Markdown file.

If you look at the raw `memes.md` file in the book repository, it contains documentation with the following annotations:

memes.md

```
<!-- semconv memes.meme -->
...
<!-- endsemconv -->
```

`https://github.com/PacktPublishing/Modern-Distributed-Trac-ing-in-.NET/blob/main/chapter14/semconv/memes.md`

Content in between these lines is auto-generated from the YAML group that has the `memes.meme` identifier. We can regenerate this content with the following command (make sure to specify the path):

```
chapter14$ docker run \
  -v /path/to/chapter14/semconv:/source \
  -v /path/to/chapter14/semconv:/destination \
  otel/semconvgen:latest \
  -f /source markdown -md /destination
```

Here, we use the Markdown generator from the `otel/semconvgen` image. We mount the `source` and `destination` volumes. The generator recursively parses all YAML files found in the `source` folder and then generates attribute tables in Markdown files available in the `destination` folder based on the `semconv` annotations we saw earlier.

The generation is the cherry on top of the cake, and you might not need it initially. Still, if you decide to use OpenTelemetry semantic convention schemas, make sure to use the `otel/semconvgen` tool to *validate* YAML files, which you can do as a part of your CI run; just add the `-md-check` flag to the previous command.

The tooling also supports generating attribute definitions in code using Jinja templates (`https://jinja.palletsprojects.com`). All we need is to create a Jinja template for the `SemanticConventions.cs` file and run the `otel/semconvgen` generator.

We can also define tracing, metrics, or event-specific conventions. Let's do it for events.

Defining event conventions

Meme upload and download events are important for business reporting. We can't really expose them as metrics – meme names have high cardinality and we're interested in finding the most popular ones or measuring other per-meme things.

To avoid breaking business reporting, we need to make sure events are defined precisely enough and are well documented. To achieve this, we can declare an event in the following way:

---

memes-events.yaml

```
- id: meme.download.event
  type: event
  prefix: download_meme
  brief: "Describes meme download event."
  Attributes:
    - ref: memes.meme.name
    - ref: memes.meme.size
      requirement_level: required
    - ref: memes.meme.type
      requirement_level: required
```

https://github.com/PacktPublishing/Modern-Distributed-Tracing-in-.NET/blob/main/chapter14/semconv/memes-events.yaml

---

Here, we declare a group with the `event` type (previously, we used `attribute_group`, which is signal agnostic). We provided a prefix (`download_name`) that documents the event name. We added references to attributes defined previously, but now require the presence of the `size` and `type` attributes on these events only.

You might have noticed that the event name does not contain a namespace – here, we follow the `event` semantic convention. If you look at the corresponding code snippet for the `EventService` class, we also record the `event.domain` attribute, which serves as a namespace.

With this approach, we can define spans or metric conventions that reuse the same common attributes.

These schemas or corresponding Markdown files would define and document the contract between telemetry producers and consumers regardless of the language they use.

Summary

In this chapter, we talked about different ways to keep custom telemetry and attributes consistent across your system. We identified attribute properties to be documented and learned about attribute naming conventions.

Keeping telemetry consistent is a challenge. We explored how to make it easier by sharing common instrumentation code, including OpenTelemetry setup and utility methods that report attributes with the right names and types.

Finally, we learned about the OpenTelemetry semantic conventions schema and tooling, which may help you define, validate, and automate the documentation process for custom conventions.

Defining a common schema for telemetry during the early stages of a project is going to save your organization a lot of time down the road, and now you have the knowledge and tools to do it. In the next chapter, we'll talk about brownfield systems, where new solutions coexist with legacy ones, and we'll see how difficult it can be to align different standards and conventions.

Questions

1. It's likely that alerts, dashboards, and usage reporting depend on custom telemetry and rely on conventions. How would you approach evolving conventions to prevent breaking something critical?

2. Is it possible to validate that telemetry coming from some service follows defined semantic conventions?

15

Instrumenting Brownfield Applications

When building brand-new services and systems, it's easy to achieve a basic level of observability with distributed traces, metrics, and logs using OpenTelemetry instrumentation libraries.

However, we don't usually create applications from scratch – instead, we evolve existing systems that include services in different stages of their life, varying from experimental to legacy ones that are too risky to change.

Such systems normally have some monitoring solutions in place, with custom correlation formats, telemetry schemas, logs and metrics management systems, dashboards, alerts, as well as documentation and processes around these tools.

In this chapter, we'll explore instrumentation options for such heterogeneous systems, which are frequently referred to as **brownfield**. First, we'll discuss instrumentation options for legacy parts of the system and then look deeper into context propagation and interoperating with legacy correlation formats. Finally, we'll talk about existing monitoring solutions and investigate migration strategies.

You'll learn to do the following:

- Pick a reasonable level of instrumentation for legacy services
- Leverage legacy correlation formats or propagate context transparently to enable end-to-end tracing
- Forward telemetry from legacy services to new observability backends

By the end of this chapter, you will be able to implement distributed tracing in your brownfield application, keeping changes to legacy parts of a system to a minimum.

Technical requirements

The code for this chapter is available in the book's repository on GitHub at `https://github.com/PacktPublishing/Modern-Distributed-Tracing-in-.NET/tree/main/chapter15`.

To run samples for this chapter, we'll need a Windows machine with the following tools:

- .NET SDK 7.0 or later
- .NET SDK 4.6.2
- Docker and `docker-compose`

Instrumenting legacy services

The word **legacy** has a negative connotation in software development, implying something out of date and not exciting to work on. In this section, we will focus on a different aspect and define a legacy service as something that mostly successfully does its job but no longer evolves. Such services may still receive security updates or fixes for critical issues, but they don't get new features, refactoring, or optimizations.

Maintaining such a service requires a different set of skills and fewer people than the evolving one, so the context of a specific system can easily get lost, especially after the team that was developing it moved on and now works on something else.

As a result, changing such components is very risky, even when it comes to updating runtime or dependency versions. Any modification might wake up dormant issues, slightly change performance, causing new race conditions or deadlocks. The main problem here is that with limited resources and a lack of context, nobody might know how a service works, or how to investigate and fix such issues. There also may no longer be appropriate test infrastructure to validate changes.

From an observability standpoint, such components usually have some level of monitoring in place, which is likely to be sufficient for maintenance purposes.

Essentially, when working on the observability of a system, we would touch legacy services only when it's critical for newer parts of the system.

Let's look at a couple of examples to better understand when changing legacy service is important and how we can minimize the risks.

Legacy service as a leaf node

Let's assume we're building new parts of the system using a few legacy services as a dependency, as shown in *Figure 15.1*:

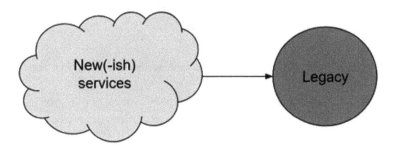

Figure 15.1 – New services depend on legacy ones

For the purposes of our new observability solution, we may be able to treat a legacy system as a black box. We can trace client calls to the legacy components and measure client-side latency and other stats. Sometimes, we'll need to know what happens inside the legacy component – for example, to understand client-side issues or work around legacy system limitations. For this, we can leverage existing logging and monitoring tools available in the legacy services. It could be inconvenient, but if it is rare, it can be a reasonable option.

If legacy components support any correlation headers for incoming requests, we can populate them on the client side to correlate across different parts of a system. We'll look at this in the *Propagating context* section of this chapter.

Another thing we may be able to do without changing a legacy system is forking and forwarding its telemetry to the same observability backend – we'll take a closer look at this in the *Consolidating telemetry from legacy-monitoring tools* section.

Being able to correlate telemetry from new and legacy components and store it in the same place could be enough to debug occasional integration issues.

Things get more interesting if a legacy system is in the middle of our application – let's see why.

A legacy service in the middle

When we refactor a distributed system, we can update downstream and upstream services around a legacy component, as shown in *Figure 15.2*:

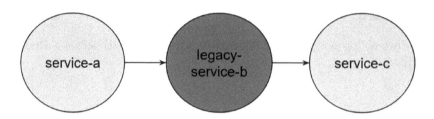

Figure 15.2 – Legacy service-b is in between the newer service-a and service-c

From the tracing side, the challenge here is that the legacy component does not propagate W3C Trace Context. Operations that go through **legacy-service-b** are recorded as two traces – one started by **service-a** and another started by **service-c**.

We need to either support legacy context propagation format in newer parts of the system, or update the legacy component itself to enable context propagation.

Before we go into the context propagation details, let's discuss the appropriate level of changes we should consider applying to a service, depending on the level of its maturity.

Choosing a reasonable level of instrumentation

Finding the right level of instrumentation for mature parts of a system depends on how big of a change is needed and how risky it is. Here are several things to consider:

- Where do legacy services send telemetry to? Is it the same observability backend that we want to use for the newer parts?

- How critical is it for the observability of the overall system to get telemetry from legacy components?

- Do legacy services support some context propagation format? Can we interoperate with it from newer services?

- Can we change some of our legacy services? How old is the .NET runtime? Do we have an adequate testing infrastructure? How big is the load on this service? How critical is the component?

Let's go through a few solutions that may apply, depending on your answers.

Not changing legacy services

When legacy parts of a system are instrumented with a vendor-specific SDK or agent and send telemetry to the same observability backend as we want to use for newer parts, we might not need to do anything – correlation might work out of the box or with a little context propagation adapter in newer parts of the system.

Your vendor might have a migration plan and documentation explaining how to make services, using their old SDK and OpenTelemetry-based solution, produce consistent telemetry.

Another case when doing nothing is a good option is when our legacy components are mostly isolated and either work side by side with newer parts or are leaf nodes, as shown in *Figure 15.1*. Then, we can usually develop and debug new components without data from legacy services.

We could also be able to tolerate having broken traces, especially if they don't affect critical flows and we're going to retire legacy services soon.

Doing nothing is the best, but if it's problematic for overall observability, the next discreet option is passing context though a legacy system.

Propagating context only

If newer parts communicate with legacy services back and forth and we can't make trace context propagation work, it can prevent us from tracing critical operations through a system. The least invasive change we can do then is to transparently propagate trace context through a legacy service.

When such a service receives a request, we would read the trace context in W3C (B3, or another format) and then pass it through, without any modification to all downstream services.

This way, legacy services will not appear on traces, but we will have consistent end-to-end traces for the newer parts.

We can possibly go further and stamp trace context on the legacy telemetry to simplify debugging.

If transparent context propagation is still not enough and we need to have telemetry from all services in one place, the next option to consider is forking legacy telemetry and sending it to the new observability backend.

Forwarding legacy telemetry to the new observability backend

Debugging issues across different observability backends and log management tools can be challenging, even when data is correlated.

To improve it, we may be able to intercept telemetry from the legacy system on the way to its backend or enable continuous export from that backend to the new one used by the rest of the system.

Forwarding may require configuration changes on the legacy system, and even if such changes are small, there is still a risk of slowing down the telemetry pipeline and causing an incident for the legacy service.

The younger and the more flexible the system is, the more changes we can consider, and the most invasive one is onboarding a legacy system onto OpenTelemetry and enabling network instrumentations.

Adding network-level instrumentation

It's likely that legacy telemetry is not consistent with distributed traces coming from new services. We may be able to transform it, or can sometimes tolerate the difference, but we may as well consider enabling minimalistic distributed tracing in legacy services. This will take care of context propagation and produce consistent telemetry with the rest of the system.

With this approach, we'll pump new telemetry from legacy services to the new backend and keep all existing instrumentations and pipelines running to avoid breaking existing reports, dashboards, and alerts.

Something to be aware of here is that OpenTelemetry works on .NET 4.6.2 or newer versions of .NET. While instrumentations for IIS, classic ASP.NET, and OWIN are available in the **contrib** repository (at `https://github.com/open-telemetry/opentelemetry-dotnet-contrib`), such instrumentations do not get as much love as newer ones.

You might also hit some edge cases with `Activity.Current` when using IIS – it can get lost during hopping between managed and native threads.

Onboarding existing services to OpenTelemetry while keeping old tools working can be a first step in a migration project, which eventually sunsets legacy monitoring solutions.

This is a viable solution for any mature service and should be considered unless the service is on a retirement path already. However, if it's not an option, we can still combine and evolve other approaches mentioned here. Let's now look at the practical side and see how we can do it.

Propagating context

The first goal for context propagation is to enable end-to-end distributed tracing for new services, even when they communicate through legacy ones, as shown in *Figure 15.2*. As a stretch goal, we can also try to correlate telemetry from new and legacy parts.

The solution that would work in most cases involves enabling context propagation in legacy services. Depending on how legacy services are implemented, this change can be significant and risky. So, before we do it, let's check whether we can avoid it.

Leveraging existing correlation formats

Our legacy services might already propagate context, just in a different format. One popular approach is to pass a correlation ID that serves the same purpose as a trace ID in the W3C Trace Context standard, identifying a logical end-to-end operation.

While correlation ID is not compatible with trace context out of the box, it may be possible to translate one to another.

In a simple case, correlation ID is just a string, and then we just need to pass it to the legacy service in a header. Then, we can expect it to propagate it as is to downstream calls, as shown in *Figure 15.3*:

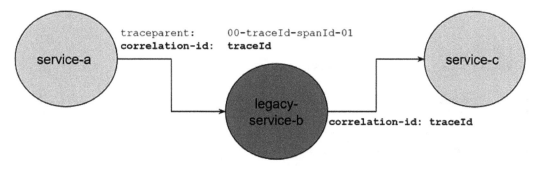

Figure 15.3 – Passing the W3C Trace ID via a legacy correlation header

Here, **service-a** populates the `correlation-id` header along with `traceparent`, **legacy-service-B** picks `correlation-id` up, ignoring the unknown `traceparent`, and passes it over to **service-c**. In turn, **service-c** supports both the `traceparent` and `correlation-id` values. It only has `correlation-id`, so it uses it to continue the trace started by **service-a**.

Let's implement it with a custom OpenTelemetry context propagator, starting with the injection side, as shown in the following code snippet:

CorrelationIdPropagator.cs

```
public override void Inject<T>(PropagationContext context,
  T carrier, Action<T, string, string> setter)
{
  if (context.ActivityContext.IsValid())
    setter.Invoke(carrier,
      CorrelationIdHeaderName,
      context.ActivityContext.TraceId.ToString());
}
```

https://github.com/PacktPublishing/Modern-Distributed-Tracing-in-.NET/blob/main/chapter15/Brownfield.OpenTelemetry.Common/CorrelationIdPropagator.cs

Here, we check whether the activity context is valid and set `TraceId` as a string on the `correlation-id` header. We're setting this propagator up to run after the `TraceContextPropagator` implementation available in OpenTelemetry, so there is no need to take care of Trace Context headers here.

And here's the extraction code:

CorrelationIdPropagator.cs

```
public override PropagationContext Extract<T>(
  PropagationContext context, T carrier,
  Func<T, string, IEnumerable<string>> getter)
{
  if (context.ActivityContext.IsValid()) return context;

  var correlationIds = getter.Invoke(carrier,
   CorrelationIdHeaderName);

  if (TryGetTraceId(correlationIds, out var traceId))
  {
    var traceContext = new ActivityContext(
      ActivityTraceId.CreateFromString(traceId),
      ActivitySpanId.CreateRandom(),
      ActivityTraceFlags.Recorded,
      isRemote: true);
    return new PropagationContext(traceContext,
      context.Baggage);
  }
  ...
}
```

https://github.com/PacktPublishing/Modern-Distributed-Trac-
ing-in-.NET/blob/main/chapter15/Brownfield.OpenTelemetry.Common/
CorrelationIdPropagator.cs

The custom extraction we implemented here runs after trace context extraction, so if there was a valid `traceparent` header in the incoming request, then `context.ActivityContext` is populated by the time the `Extract` method is called. Here, we give priority to W3C Trace Context and ignore the `correlation-id` value.

If `context.ActivityContext` is not populated, we retrieve the `correlation-id` value and try to translate it to a trace ID. If we can do it, then we create a new `ActivityContext` instance, using `correlation-id` as a trace ID and a fake parent span ID.

Here's the implementation of the `TryGetTraceId` method:

CorrelationIdPropagator.cs

```
traceId = correlationId.Replace("-", "");
if (correlationId.Length < 32)
  traceId = correlationId.PadRight(32, '0');
else if (traceId.Length > 32)
  traceId = correlationId.Substring(0, 32);
```

https://github.com/PacktPublishing/Modern-Distributed-Trac-
ing-in-.NET/blob/main/chapter15/Brownfield.OpenTelemetry.Common/
CorrelationIdPropagator.cs

In this snippet, we support a variety of possible `correlation-id` formats – we remove dashes if it's a GUID, and pad or trim it if the length is not right.

> **Note**
>
> In a more complicated case, we may need to do other transformations during context extraction and injection. For example, when a legacy system requires a GUID, we can add dashes. Alternatively, if it wants a `base64`-encoded string, we can decode and encode the trace ID.

Let's now check out the traces we get with this approach.

First, run new parts of the system with the `$ docker-compose up --build` command. It starts with **service-a**, **service-c**, and the observability stack.

We also need to start **legacy-service-b**, which is the .NET Framework 4.6.2 application running on Windows. You can start it with your IDE or the following command:

```
legacy-service-b$ dotnet run --correlation-mode correlation-id
```

Then, hit the following URL in your browser: `http://localhost:5051/a?to=c`. This will send a request to **service-a**, which will call **service-c** through **legacy-service-b**.

Now, let's open Jaeger at `http://localhost:16686` and find the trace from **service-a**, which should look like the one shown in *Figure 15.4*:

Figure 15.4 – An end-to-end trace covering service-a and service-c

As you can see, there is no **legacy-service-b** in the figure – it does not send telemetry to Jaeger. The only indication is the endpoint – the port (`5050`) belongs to **legacy-service-b**.

There is just one trace, but it still looks broken – spans are correlated, but parent-child relationships between the client span on **service-a** and the server span on **service-c** are lost.

Still, it's an improvement. Let's now disable the `correlation-id` support on **service-a** and **service-c**. We can do it by changing the `Compatibility__SupportLegacyCorrelation` environment variable in `docker-compose.yml` to `false` on both services and restarting the docker compose application. Then, we'll see two independent traces for **service-a** and **service-c**, so even the correlation will be lost.

> **Note**
>
> By relying on the existing context propagation format and implementing a custom propagation adapter, we can usually record end-to-end traces for new services without any modification to the legacy ones.

Can we also correlate telemetry from the legacy and new services? Usually, legacy services stamp their version of `correlation-id` on all logs. If that's the case, we can search using the trace ID across all telemetry but may need to map the trace ID to the correlation ID and back, in the same way we did with the propagator.

However, what if we didn't have custom correlation implemented in a legacy service or were not able to implement an adapter? We'd need to modify the legacy service to enable context propagation – let's see how it can be done.

Passing context through a legacy service

Essentially, if there is no existing context propagation mechanism, we can implement one. To minimize changes to legacy systems, we can propagate context transparently, without modifying it.

We need to intercept incoming and outgoing requests to extract and inject trace context, and we also need a way to pass the context inside the process.

The implementation of this approach, especially the interception, depends on the technologies, libraries, and patterns used in a specific legacy service.

Incoming request interception can be achieved with some middleware or request filter. If IIS is used, it can be also done in a custom HTTP telemetry module, but then we cannot fully rely on ambient context propagation due to managed-to-native thread hops.

Passing context within a process can be usually achieved with `AsyncLocal` on .NET 4.6+ or `LogicalCallContext` on .NET 4.5 – this way, it will be contained in the new code and won't require plumbing context explicitly.

In our demo system, **legacy-service-b** is a self-hosted OWIN application, and we can implement context extraction in the OWIN middleware:

PassThroughMiddleware.cs

```
private static readonly
  AsyncLocal<IDictionary<string, object>> _currentContext =
    new AsyncLocal<IDictionary<string, object>>();

public static IDictionary<string, object> CurrentContext =>
  _currentContext.Value;

public override async Task Invoke(IOwinContext context)
{
  var tc = EmptyContext;
  if (context.Request.Headers.TryGetValue("traceparent",
    out var traceparent))
  {
    tc = new Dictionary<string, object>
      {{ "traceparent", traceparent[0] }};
    ...
  }
  _currentContext.Value = tc;
  ...

  using (var scope = _logger.BeginScope(tc))
  {
    await Next.Invoke(context);
  }
}
```

https://github.com/PacktPublishing/Modern-Distributed-Trac-ing-in-.NET/blob/main/chapter15/legacy-service-b/PassThrough/PassThroughMiddleware.cs

First, we declare a static AsyncLocal value that holds trace context, represented with a simple dictionary.

In the middleware Invoke method, we read traceparent along with the tracestate and baggage headers (which are omitted for brevity). We populate them in the trace context dictionary. Depending on your needs, you can always limit supported context fields to traceparent only and optimize the code further.

Then, we populate the context dictionary on the _currentContext field, which we can then access through the public CurrentContext static property.

The last thing we do here is to invoke the next middleware, which we wrap with a logger scope containing the context dictionary. This allows us to populate trace context on all logs coming from **legacy-service-b**, thus correlating them with telemetry coming from new services.

In practice, legacy applications rarely use `ILogger`, but logging libraries usually have some other mechanism to populate ambient context on log records. Depending on the library, you may be able to access and populate `CurrentContext` with little change to the logging configuration code.

Getting back to context propagation, we now need to inject the `CurrentContext` value into the outgoing requests.

In the case of HTTP and when .NET `HttpClient` is used, we can do it with custom `DelegatingHandler` implementation. It will be more tedious with `WebRequest` usage spread across the application code when there are no helper methods that create them consistently.

The handler implementation is shown in the following code snippet:

PassThroughHandler.cs

```
protected override Task<HttpResponseMessage> SendAsync(
  HttpRequestMessage request, CancellationToken token)
{
  foreach (var kvp in PassThroughMiddleware.CurrentContext)
    request.Headers.Add(kvp.Key, pair.Value?.ToString());

  return base.SendAsync(request, token);
}
```

https://github.com/PacktPublishing/Modern-Distributed-Trac-ing-in-.NET/blob/main/chapter15/legacy-service-b/PassThrough/PassThroughMiddleware.cs

Here, we just inject all fields from `CurrentContext` on outgoing request headers and then invoke the next handler. That's it.

> **Note**
>
> Starting with the `System.Diagnostics.DiagnosticSource` package version 6.0.0, .NET provides a `DistributedContextPropagator` base class along with several implementations, including W3C trace context and a pass-through propagator. It can be useful if you can add a dependency on a newish `DiagnosticSource` package, or when configuring propagation for native distributed tracing instrumentations in ASP.NET Core and `HttpClient`. In the case of our legacy service, extraction and injection alone are trivial, so adding a new dependency is not really justified.

Now, we can run the application again and check the traces:

1. Start new services with $ `docker-compose up --build` and then **legacy-service-b** with the following command:

    ```
    legacy-service-b$ dotnet run --correlation-mode pass-through
    ```

2. Then call **service-a** with `http://localhost:5051/a?to=c` again and open Jaeger. We should see a trace like the one in *Figure 15.5*:

Figure 15.5 – An end-to-end trace with transparent service-b

Here, we have correlation and causation – the client span on **service-a** is a direct parent of the server span on **service-c**. However, **service-b** is nowhere to be seen, as it does not actively participate in the tracing.

Now, we have a couple of options to pass context through the legacy system, but we can be creative and come up with more options specific to our application – for example, we can stamp legacy correlation or request IDs on the new telemetry, or log them and then post-process telemetry to correlate broken traces.

With these options, we should be able to achieve at least some level of correlation. Let's now check how we can forward telemetry from legacy services to the new observability backends.

Consolidating telemetry from legacy monitoring tools

One of the biggest benefits a good observability solution can provide is low cognitive load when debugging an application and reading through telemetry. Even perfectly correlated and high-quality telemetry is very hard to use if it's spread across multiple tools and can't be visualized and analyzed together.

When re-instrumenting legacy services with OpenTelemetry is not an option, we should check whether it's possible to forward existing data from legacy services to a new observability backend.

As with context propagation, we can be creative and should start by leveraging existing solutions. For example, old .NET systems usually report and consume Windows performance counters and send logs to EventLog, or store them on the hard drive.

The OpenTelemetry Collector provides support for such cases via receivers, available in the contrib repository (at `https://github.com/open-telemetry/opentelemetry-collector-contrib`).

For example, we can configure a file receiver with the following snippet:

otel-collector-config.yml

```yml
filelog:
  include: [ /var/log/chapter15*.log ]
  operators:
    - type: json_parser
      timestamp:
        parse_from: attributes.Timestamp
        layout: '%Y-%m-%dT%H:%M:%S.%f'
      severity:
        parse_from: attributes.LogLevel
```

https://github.com/PacktPublishing/Modern-Distributed-Trac-ing-in-.NET/blob/main/chapter15/configs/otel-collector-config.yml

Here, we configure the collector receiver and specify the log file location and name pattern. We also configure mapping and transformation rules for individual properties in log records. In this example, we only map timestamp and log level, but if log records are structured, it's possible to parse other properties using similar operators.

We can also rely on our backend to grok unstructured log records or parse records at a query time if we rarely need the data.

Here's an example of collector output with a parsed log record, which, depending on your collector configuration, can send logs to the new observability backend:

```
Timestamp: 2023-05-27 01:00:41.074 +0000 UTC
SeverityText: Information
...
Attributes:
     -> Scopes: Slice([{"Message":"System.Collections.Generic.
Dictionary`2[System.String,System.Object]","traceparent":"00-
78987df9861c2d7e46c10bd084570122-53106042475b3e32-01"}])
     -> Category: Str(LegacyServiceB.LoggingHandler)
...
     -> State: Map({"Message":"Request complete. GET http://
localhost:5049/c, OK","method":"GET","status":"OK","url":"http://
localhost:5049/c","{OriginalFormat}":"Request complete. {method}
{url}, {status}"})
Trace ID:
Span ID:
```

As you can see, we could also configure the receiver to parse the `traceparent` value populated in the log scopes to record `Trace ID` and `Span ID` for the proper correlation.

You can reproduce it by running **legacy-service-b** with the following command and sending some requests to it directly, or via **service-a**:

```
legacy-service-b $ dotnet run --correlation-mode pass-through > ../
tmp/logs/chapter15.log
```

A collector can be helpful in sidecar mode, forwarding data available on the machine where legacy service instances are running, and collecting performance counters or logs. It can also pretend to be our old backend and receive Zipkin or Jaeger spans, Prometheus metrics, and vendor-specific signals, such as Splunk metrics and logs.

We can write custom receivers and leverage collector transformation processors to produce consistent telemetry whenever possible.

In addition to the endless possibilities a OpenTelemetry Collector can provide, we should check whether the observability vendor we use for legacy services allows continuous export for collected telemetry, which would allow us to get the data without changing anything on the legacy system.

Summary

In this chapter, we explored tracing in brownfield applications, where some of the services can be hard to change and onboard onto a full-fledged observability solution with OpenTelemetry.

We discussed possible levels of instrumentation for such services and found several cases when we can avoid changing old components altogether. Then, we went through the changes we can apply, starting with minimalistic transparent context propagation and going all the way to onboarding onto OpenTelemetry.

Finally, we applied some of these options in practice, enabling end-to-end correlation through a legacy service and forwarding file logs to the OpenTelemetry Collector.

Now, you should be ready to come up with the strategy for your own legacy components and have the building blocks to implement it.

This chapter concludes our journey into distributed tracing and observability on .NET – I hope you enjoyed it! The observability area is evolving fast, but now you have a foundational knowledge to design and implement your systems with observability in mind, evolve them by relying on relevant telemetry data, and operate them with more confidence, knowing what telemetry represents and how it's collected. Now, it's time to apply your knowledge or create something new based on it.

Questions

1. How would you approach instrumenting an existing service that is a critical part of most user scenarios in your system? This service is mature and is rarely changed, but there are no plans to retire it any time soon.

2. What can go wrong when we add OpenTelemetry to a legacy service?

3. When implementing transparent context propagation, can we leverage the `Activity` class instead of adding our own context primitive and the `AsyncLocal` field?

Assessments

Chapter 1 – Observability Needs of Modern Applications

1. You can think about a span as a structured event with a strict but extensible schema, allowing you to track any interesting operation. Spans have trace context that describes the relationships between them. They also have a name, start time, end time, status, and a property bag, with attributes to represent operation details.

    Complex and distributed operations need multiple spans that describe at least each incoming and outgoing request. A group of such correlated spans that share the same `trace-id` is called a trace.

2. Spans (also known as Activities in .NET) are created by many libraries and applications. To enable correlation, we need to propagate context within the process and between processes.

    In .NET, we use `Activity.Current` to propagate context within the process. This is a current span that flows with an execution context in synchronous or asynchronous calls. Whenever a new activity is started, it uses `Activity.Current` as its parent and then becomes current itself.

    To propagate the trace context between the processes, we pass it over the wire to the next service. W3C Trace Context is a standard propagation format for the HTTP protocol, but some services use the B3 format.

3. There is no single answer to this question, but here're some general considerations on how you can leverage a combination of signals coming from your service:

    *   Check whether the problem is widespread and affects more than this user and request. Is your service healthy overall? Is it specific to the API path the user hits, region, partition, feature flag, or new service version? Your observability backend might be able to assist with it

    *   If the problem is not widespread, find traces for problematic requests using trace context if it is known, or filtering by known attributes. If you see gaps in traces, retrieve logs for this operation. If that's not enough, use profiling to investigate further. Consider adding more telemetry.

    *   For widespread issues, you might find the root cause of the problem by identifying specific attributes correlated with the reported problem.

    *   Otherwise, narrow down the issue layer by layer. Are dependencies working fine? Is there something new upstream? Any changes in the load?

- If issues are not specific to any combination of attributes, check the dependency health and resource utilization. Check the crash and restart count, CPU load, memory utilization, extensive garbage collection, I/O, and network bottlenecks.

Chapter 2 – Native Monitoring in .NET

1. Use `Activity.Current?.Id` on the page. For example, like this: `<p>traceparent: <code>@System.Diagnostics.Activity.Current?.Id</code></p>`.

2. If we have `dotnet-monitor` running as a sidecar, we can connect to its instance corresponding to the problematic service instance, check the metrics and logs, and create dumps. We could even configure `dotnet-monitor` to trigger a dump collection based on certain events or resource consumption thresholds.

    If we don't have `dotnet-monitor`, but can access service instances, we can install `dotnet-monitor` there and get diagnostics information from the running process.

    If instances are healthy, but the problem is somewhere inside the telemetry pipeline, troubleshooting steps would depend on the tools we use. For example, with Jaeger we can check logs; the Prometheus UI shows connectivity with targets; the OpenTelemetry collector provides logs and metrics for self-diagnostics.

3. Query:

    ```
    sum by (service_name, http_route)
        (rate(http_server_duration_ms_count[1m]))
    ```

    The query sums up the request rates across all running service instances, grouping it by service name and `http_route` (which represents the API route).

    The rate function (`rate(http_server_duration_ms_count)` first calculates the rate per second, then averages the rate over one minute.

4. Search the traces with the URL and method filter in Jaeger. For uploads, it would be `http.url=http://storage:5050/memes/<name>` `http.method=PUT`. To find downloads, we would use `http.url=http://storage:5050/memes/<name>` `http.method=GET`. However, this isn't convenient and we should consider adding the meme name as an attribute on all spans.

Chapter 3 – The .NET Observability Ecosystem

1. Check the registry (`https://opentelemetry.io/registry/`) and OpenTelemetry .NET repo. If you don't see your library in any of them, search in issues and PRs. It's also a good idea to search whether anything is available in the library GitHub repo or documentation.

When you find an instrumentation, there are several things to check for:

- **Version and stability**: Beta instrumentations could still have a high quality and be battle-tested but do not guarantee API or telemetry stability

- **Performance and thread safety**: Understanding the mechanism behind instrumentation is important to identify possible limitations and issues in advance

2. The most common way to instrument libraries and frameworks is `ActivitySource`—it's the .NET analog of OpenTelemetry Tracer, which can start activities. You can configure OpenTelemetry to listen to a source by its name. You might also see instrumentations using `DiagnosticSource`—it's an older and less structured mechanism available in .NET.

   It's also common to leverage hooks provided by libraries that can be global or applied to specific instances of the client.

3. Service meshes can trace requests to and from service mesh sidecars and provide insights into retries, service discovery, or load balancing. If they handle communication with cloud service, remote database, or queue, they can instrument corresponding communication. Service meshes can propagate the context from one application to another but cannot propagate it within the service from incoming to outgoing calls.

Chapter 4 – Low-Level Performance Analysis with Diagnostic Tools

1. If your service defines SLIs, check them first and see whether they are within the boundaries defined by your SLOs. In other words, check the key metrics that measure your user experience and see whether they are within healthy limits. For REST API-based services, it is usually the throughput of successful requests and latency grouped by API and other things that are important in your application.

   Resource consumption metrics could be correlated to user experience, but do not determine it. They (and other metrics that describe the internals of your service) can help you understand why the user experience has degraded and can predict future issues with some level of confidence.

2. First, we should try to find which service is responsible: check upstream and downstream services for whether the load on your service is normal and properly distributed across instances. Check whether dependencies are healthy using their server-side metrics when possible.

   If we can narrow down the issue to a specific service, we can check whether the issue is specific to a certain instance or group of instances, or whether instances are restarting a lot. For affected instances, we can check their resource utilization patterns for memory, CPU, GC frequency, threads, contentions, or anything that looks unusually high or low. Then, we can capture a dump from the problematic instance(s) to analyze memory and thread stacks.

3. Performance tracing (also known as profiling or just tracing) is a technique that allows us to capture detailed diagnostics about application behavior and code – call stacks, GC, contention, network events, or anything else that .NET or third-party libraries want to expose. Such events are off by default but can be enabled and controlled inside the process and out-of-process. Tools such as `dotnet-trace`, `dotnet-monitor`, PerfView, PerfCollect, JetBrains dotTrace, Visual Studio, and continuous profilers can collect and visualize them. Performance tracing can be used to investigate functional and performance issues or optimize your code.

Chapter 5 – Configuration and Control Plane

1. We'd need tail-based sampling that's applied after span or trace ends and we know the duration or if there were any failures. Tail-based sampling can't be done inside the process since we have distributed multi-instance applications, but we can use a tail-based sampling processor in the OpenTelemetry Collector that buffers traces and then samples them based on latency, or status codes.

    If we only capture suspicious traces, we will not have a baseline anymore – we won't be able to use traces to observe normal system behavior, build analytics, and so on. So, we should additionally capture a percentage or rate of random traces – if we mark them somehow, we can analyze them separately from problematic traces to create unbiased analytics.

    It's always a good idea to rate-limit all traces, so we don't overload the telemetry pipeline with traffic bursts.

    In addition to sampling configuration on the OpenTelemetry Collector, we should consider configuring probability sampling on individual .NET services – depending on this, we would allocate an appropriate number of resources for Collector and also balance the performance impact of the instrumentation.

2. Let's record a try number using the OpenTelemetry `http.resend_count` attribute that should be set on each HTTP span that represents a retry or redirect. We can use the `EnrichWithHttpRequestMessage` hook on the HTTP client instrumentation to intercept the outgoing request and its activity, but where would we get the retry number from?

    Well, we can maintain it in our retry handler (if you use Polly, you could use `Context` instead) and pass it to the hook via `HttpRequestMessage.Options`. So, the final solution could look like this:

Program.cs

```
AddHttpClientInstrumentation(options =>
{
options.EnrichWithHttpRequestMessage = (act, req) =>
{
if (req.Options.TryGetValue(
```

```
new HttpRequestOptionsKey<int>("try"),
out var tryCount) && tryCount > 0)
act.SetTag("http.resend_count", tryCount);
...
}
}
```

https://github.com/PacktPublishing/Modern-Distributed-Trac-ing-in-.NET/blob/main/chapter5/memes/frontend/Program.cs

RetryHandler.cs

```
for (int i = 0; i < MaxTryCount; i++)
{
request.Options.Set(new
HttpRequestOptionsKey<int>("try"), i);
try
{
var response = await base.SendAsync(request,
token);
...
}
catch (Exception e) { ... }
await Task.Delay(delays[i]);
}
```

https://github.com/PacktPublishing/Modern-Distributed-Trac-ing-in-.NET/blob/main/chapter5/memes/frontend/RetryHandler.cs

3. Let's check out the OpenTelemetry Collector documentation for tail-based sampling at https://github.com/open-telemetry/opentelemetry-collector-contrib/blob/main/processor/tailsamplingprocessor/README.md. We need to declare and configure the `tail_sampling` processor and add it to the pipeline. Here's a sample configuration:

otel-collector-config.yml

```
processors:
...
tail_sampling:
decision_wait: 2s
expected_new_traces_per_sec: 500
policies:
[{ name: limit-rate,
```

```
type: rate_limiting,
rate_limiting: {spans_per_second: 50}}]
service:
pipelines:
traces:
receivers: [otlp]
processors: [tail_sampling, batch]
exporters: [jaeger]
```

https://github.com/PacktPublishing/Modern-Distributed-Trac-
ing-in-.NET/blob/main/chapter5/memes/configs/otel-collec-
tor-config.yml

You can check your current rate of recorded spans using the `rate(otelcol_receiver_accepted_spans[1m])` query in Prometheus and monitor the exported rate with the `rate(otelcol_exporter_sent_spans[1m])` query.

Chapter 6 – Tracing Your Code

1. When setting up OpenTelemetry, you can enable `ActivitySource` by calling into the `TracerProviderBuilder.AddSource` method and passing the source name. OpenTelemetry will then create an `ActivityListener` – a low-level .NET API that listens to `ActivitySource` instances. The listener samples activities using the callback provided by OpenTelemetry and notifies OpenTelemetry when activities start or end.

2. Activity (or span) events can be used to represent something that happens at a point in time or is too short to be a span and does not need individual context. At the same time, events must happen in the scope of some activity and are recorded along with it. Activity events stay in memory until the activity is garbage-collected and their number is limited on the exporter side.

    Logs are usually a better alternative to `Activity` events as they are not necessarily tied to specific activity, sampling, or exporter limitations. OpenTelemetry treats events and logs similarly. Events expressed as log records are structured and can follow specific semantic conventions.

3. Links provide another way to correlate spans with cover scenarios when the span has multiple parents or is related in some way to several other spans at once. Without links, spans can only have one parent and multiple children and can't be related to spans in other traces.

    Links are used in messaging scenarios to express receiving or processing multiple independent messages at once. When we process multiple messages, we need to extract the trace context and create an `ActivityLink` from each of them. Then, we can pass a collection of these links to the `ActivitySource.StartActivity` method. We can't change these links after the corresponding `Activity` starts. Observability backends support (or don't support) links in different ways and we might need to adjust the instrumentation based on the backend capabilities.

Chapter 7 – Adding Custom Metrics

1. We should first decide what we need the metric for. For example, if we need it to rank memes in search results or to calculate ad hits, we should separate it from telemetry. Assuming we store the meme download counter in a database for business logic purposes, we could also stamp it on traces or events as an attribute when the counter is updated.

    From a telemetry-only standpoint, metric per meme would have high cardinality as we probably have millions of memes in the system and thousands active per minute. With some additional logic (for example, if we can ignore rarely accessed memes), we might even be able to introduce a metric with a meme name as an attribute.

    I would start with traces and aggregate spans by meme name in a rich query. Even if traces are sampled, I can still calculate the estimated number of downloads, compare it between memes, and see trends.

2. Usually, both, but it depends: we need incoming HTTP request traces to investigate individual failures and delays and know what normal request flow looks like under different conditions. Do we need metrics as well? Probably yes. At a high scale, we sample traces aggressively but likely need more precise data than estimated counts. Another problem is that even if we don't sample or don't mind rough estimates, querying over all spans during the time window can be expensive and long – it might need to process millions of records. If we build dashboards and alerts on this data, we want queries to be fast and cheap. Even if they are used for ad hoc analysis during incidents, we still want queries to be fast.

    So, the answer depends on the observability backend, what it is optimized for, and its pricing model, but collecting both gives us a good starting point.

3. For the number of active instances, we can report `ObservableUpDownCounter` with resource attributes that include instance information. The counter would always report 1 so that the sum of values across all instances at any given time will represent the number of active processes. This is how Kubernetes does it with `kube_node_info` or `kube_pod_info` metrics (check out `https://github.com/kubernetes/kube-state-metrics` for more information).

    Uptime can be reported in multiple ways – for example, as a gauge containing static start time (see `kube_node_created` or `kube_pod_start_time`) or as a resource attribute.

    Make sure to check whether your environment already emits anything similar or whether OpenTelemetry semantic conventions define a common way to report the metric you're interested in.

Chapter 8 – Writing Structured and Correlated Logs

1. The code uses string interpolation instead of semantic logging. A log message is formatted right away, so the `ILogger.Log` method is called underneath with the `"hello world: 43, bar"` string, without any indication that there are two arguments with specific names and values.

    If the `Information` level is disabled, string interpolation happens anyway, serializing all arguments and calculating just the message to be dropped.

    This code should be changed to `logger.LogInformation("hello world: {foo}, {bar}", 42, "bar")`.

2. We need to make sure that the usage report is built using log record properties that don't change:

    *   A log message would change a lot when new arguments are added or code is refactored.

    *   The logging category is usually based on a namespace, which might change during refactoring. We can consider passing categories explicitly as strings instead of a generic type parameter, but the better choice would be to make sure the report does not rely on logging categories. We can use event names or IDs – they have to be specified explicitly; we just need to make sure they are unique and don't change. One approach would be to declare them in a separate file and document that the usage reports rely on them.

3. Traces and logs describing HTTP requests contain similar information. Logs are more verbose, since we'd usually have human-readable text and need two records for one request (before and after it), with duplicated trace context and other scopes.

    If your application records all HTTP traces, there is no need to enable HTTP logging as well. If traces are sampled, there is a trade-off between the cost of capturing all telemetry and your ability to investigate rare issues. Many applications don't really need to capture all telemetry to efficiently investigate problems. For them, collecting sampled traces without HTTP logs would be the best option. If you have to investigate rare issues, one option would be to increase the sampling rate for traces. Recording HTTP logs instead is another option that comes with an additional cost to collect, store, retrieve, and analyze logs.

Chapter 9 – Best Practices

1. HTTP traces, potentially combined with some application-specific attributes, can help answer most questions about tiny RESTful service behavior. We can aggregate metrics from traces using OpenTelemetry Collector or at query time on the backend. We still need metrics for resource utilization though. The right questions to ask here are how much this solution costs us and whether there is the potential to reduce costs with sampling and how much we must spend to keep alerts running based on queries over traces. If it's a lot, then we should look into adding metrics. So, the answer is – yes, but it can be more cost-efficient to add other signals.

2. In an application under heavy load, every bug will happen again and again. No matter how small of a sampling rate we choose, we'll record at least some occurrences of such an issue. A high sampling rate would likely have some performance impact, but more importantly, it'll be very expensive to store all these traces. So, a small sampling rate should be the first choice.

3. Socket communication can be very frequent, so instrumenting every request with a span can create a huge overhead. A good starting point would be to identify how long a typical session lasts, and if it's within seconds or minutes, instrument a session with a span. Small requests can be recorded with metrics on a service side, or sometimes with logs/events.

   OpenTelemetry general and RPC semantic conventions should cover the necessary network attributes to represent the client and server and describe a request. We can also apply suitable RPC metrics to track duration and throughput.

Chapter 10 – Tracing Network Calls

1. Reusing existing instrumentation should be the first choice, especially if you don't have a lot of experience in both tracing and the gRPC stack. As you saw throughout this chapter, there are multiple details related to retries, the order of execution, and other tiny details that are hard to account for.

   Custom gRPC instrumentation makes sense if existing instrumentation does not satisfy your needs. For example, in our streaming experiments, we could optimize two layers of instrumentation (individual messages and gRPC calls) by merging them into one. We could also correlate requests, responses, and span events better if we knew the message types in the interceptor.

   Note that even custom instrumentations benefit from following semantic conventions and relying on common tooling and documentation.

2. In such an application, we should expect to see a very long span that describes a connection between the client and server. If we sample spans, we should customize the sampler to ensure we capture this span. Alternatively, we can just drop it and instead capture events that describe anything important that happens with the encompassing connection.

   Then, we should think about how/whether to trace individual messages. If they are very small and fast, tracing them individually could be too expensive because of a couple of concerns:

   - The first concern is message size. Trace context can be propagated frugally with the binary format, but still would require at least 26 bytes. You can be creative and come up with even more frugal format, propagating the message index instead of the span ID over the wire. The easiest solution would be to propagate context only for sampled-in messages and rely on metrics and events to see the overall picture.

   - The second concern is performance overhead. If your processing is very fast, tracing it might be too expensive. Sampling can help offset some of these costs, but you probably don't need to trace individual messages. Logs and events might give you the right level of observability, and you can correlate them with a message identifier.

Chapter 13 – Driving Change

1. Using a single backend for all signals has certain advantages. It should be easier to navigate between signals: for example, get all logs correlated with the trace, query events, and traces together with additional context, and jump from metrics to trace with exemplars. So, using a single backend would reduce cognitive load and minimize duplication in backend-related configuration and tooling.

   Using multiple backends can help reduce costs. For example, it's usually possible to store logs in a cheaper log management system, assuming you already have everything up and running for logs and metrics. But these backends don't always support traces well. Adding a new backend for traces and events only would make total sense.

   Tools such as Grafana may be able to provide a common UX on top of different backends to mitigate some of the disadvantages.

2. There are a few things that we need to do:

   • **Lock down the context propagation format**: Using W3C Baggage spec is a good default choice unless you already have something in place. It should be documented and, ideally, implemented and configured in internal common libraries shared by all services in your application.

   • **Documenting key naming patterns**: Make sure to use namespaces and define the root one for your system. It'll help filter everything else out. Document several common properties you want to put there – we want to make sure people use them and don't come up with something custom. Adding helper methods to populate them would also be great.

   • **Use common artifacts**: If you want to stamp baggage on telemetry, customize propagation, or just unify baggage keys, make sure to ship common internal libraries with these features.

3. When adding a cache, we're probably trying to reduce the load on a database and optimize the service response time. We should already have observability of service and database calls and can see whether the cache would help.

   If we roll this feature out gradually and conditionally, we need to be able to filter and compare telemetry based on feature flags, so we need to make sure they're recorded.

   Finally, we should make sure we have telemetry around the cache that will help us understand how it works, and why it did not work if it fails. Adding this telemetry along with feature code will have the biggest positive impact during development, testing, and initial iterations.

Chapter 14 – Creating Your Own Conventions

1. A possible solution is to define and document the stability level for attributes.

    For example, new conventions are always added at the alpha stability level. Once it's fully implemented and deployed, and you're mostly happy with the outcome, the convention can be graduated to beta.

    Conventions should stay in beta until someone tries to use them for alerts, reports, or dashboards. If it works fine, or after feedback is addressed, the convention becomes stable. After that, it cannot be changed in a breaking manner.

2. It should be possible to validate actual telemetry to some extent.

    For example, it should be possible to write a test processor (an in-process one or a custom collector component) that identifies specific spans, events, or metrics that should follow the convention and checks whether the conventions are applied consistently. This test processor could warn about issues found, flag unknown attributes, notify when expected signals were not received, and so on. It should be possible to run it as a part of integration testing in the CI pipeline.

    Another approach is to just do a regular audit on a random subset of production telemetry, which could also be automated.

Chapter 15 – Instrumenting Brownfield Applications

1. Such a service is a good candidate for migration to OpenTelemetry – since we still update it, there is probably a reasonable test infrastructure and the context within the team to prevent and mitigate failures. As a first option, we should consider adding OpenTelemetry with network instrumentation and then gradually migrating existing tools and processes onto the new observability solution, while evolving an OpenTelemetry-based approach.

    We can control the costs of this approach with sampling, enabling and moving only essential pieces onto OpenTelemetry. At some point, when we can rely on the new observability solution, we can remove corresponding legacy reporting.

2. It's likely that the .NET runtime version that the legacy service runs on is older than .NET 4.6.2, and then it's impossible to use OpenTelemetry. Even if a newer version of .NET Framework is used, adding new dependencies, such as `System.Diagnostics.DiagnosticSource` and the different `Microsoft.Extensions` packages that OpenTelemetry brings transitively, can cause runtime problems due to version conflicts.

    Other risks come from small changes and shifts in how an application works and its performance, waking up or amplifying dormant issues such as race conditions, deadlocks, or thread pool starvation.

3. If you can add newer versions of `System.Diagnostics.DiagnosticSource` as a dependency, then using `Activity` is an option.

    Note that the `Activity` class is available in .NET, starting with the `DiagnosticSource` package version 4.4.0 and .NET Core 3.0; however, it went through a lot of changes. Most of the functionality we covered in this book, including W3C Trace Context, was not available in the initial versions.

    With newer `DiagnosticSource` versions, by using `Activity`, we would modify trace context – instead of passing `traceparent` as is, we would create server and client spans and then pass an ancestor of the original `traceparent` to the downstream service. If the legacy service does not report spans to the common observability backend, we'll see correlated traces, but with missing parent-child relationships, as we saw in *Figure 15.4*.

    So, we need to have full-fledged distributed tracing implemented or, if no traces are reported, pass context through as is, without using `Activity` for it.

Index

U

W

www.packtpub.com

Subscribe to our online digital library for full access to over 7,000 books and videos, as well as industry leading tools to help you plan your personal development and advance your career. For more information, please visit our website.

Why subscribe?

- Spend less time learning and more time coding with practical eBooks and Videos from over 4,000 industry professionals

- Improve your learning with Skill Plans built especially for you

- Get a free eBook or video every month

- Fully searchable for easy access to vital information

- Copy and paste, print, and bookmark content

Did you know that Packt offers eBook versions of every book published, with PDF and ePub files available? You can upgrade to the eBook version at packtpub.com and as a print book customer, you are entitled to a discount on the eBook copy. Get in touch with us at customercare@packtpub.com for more details.

At www.packtpub.com, you can also read a collection of free technical articles, sign up for a range of free newsletters, and receive exclusive discounts and offers on Packt books and eBooks.

Other Books You May Enjoy

If you enjoyed this book, you may be interested in these other books by Packt:

.NET MAUI for C# Developers

Jesse Liberty, Rodrigo Juarez

ISBN: 978-1-83763-169-8

- Explore the fundamentals of creating .NET MAUI apps with Visual Studio
- Understand XAML as the key tool for building your user interface
- Obtain and display data using layout and controls
- Discover the MVVM pattern to create robust apps
- Acquire the skills for storing and retrieving persistent data
- Use unit testing to ensure your app is solid and reliable

Enterprise Application Development with C# 10 and .NET 6 - Second Edition

Ravindra Akella, Arun Kumar Tamirisa, Suneel Kumar Kunani, Bhupesh Guptha Muthiyalu

ISBN: 978-1-80323-297-3

- Design enterprise apps by making the most of the latest features of .NET 6
- Discover different layers of an app, such as the data layer, API layer, and web layer
- Explore end-to-end architecture by implementing an enterprise web app using .NET and C# 10 and deploying it on Azure
- Focus on the core concepts of web application development and implement them in .NET 6
- Integrate the new .NET 6 health and performance check APIs into your app
- Explore MAUI and build an application targeting multiple platforms - Android, iOS, and Windows

Packt is searching for authors like you

If you're interested in becoming an author for Packt, please visit `authors.packtpub.com` and apply today. We have worked with thousands of developers and tech professionals, just like you, to help them share their insight with the global tech community. You can make a general application, apply for a specific hot topic that we are recruiting an author for, or submit your own idea.

Share Your Thoughts

Now you've finished *Modern Distributed Tracing in .NET*, we'd love to hear your thoughts! Scan the QR code below to go straight to the Amazon review page for this book and share your feedback or leave a review on the site that you purchased it from.

`https://packt.link/r/1-837-63613-3`

Your review is important to us and the tech community and will help us make sure we're delivering excellent quality content.

Download a free PDF copy of this book

Thanks for purchasing this book!

Do you like to read on the go but are unable to carry your print books everywhere? Is your eBook purchase not compatible with the device of your choice?

Don't worry, now with every Packt book you get a DRM-free PDF version of that book at no cost.

Read anywhere, any place, on any device. Search, copy, and paste code from your favorite technical books directly into your application.

The perks don't stop there, you can get exclusive access to discounts, newsletters, and great free content in your inbox daily

Follow these simple steps to get the benefits:

1. Scan the QR code or visit the link below

https://packt.link/free-ebook/9781837636136

2. Submit your proof of purchase
3. That's it! We'll send your free PDF and other benefits to your email directly

www.ingramcontent.com/pod-product-compliance
Lightning Source LLC
Chambersburg PA
CBHW062100050326
40690CB00016B/3154